Philosophy of Engineering and Technology

Volume 47

The *Philosophy of Engineering and Technology* book series provides the multifaceted and rapidly growing discipline of the philosophy of technology with a central overarching and integrative platform. It publishes on all topics in the philosophy of technology and is open to all research communities across the world.

Specifically, it publishes edited volumes and monographs in:

- the phenomenology, anthropology and socio-politics of technology and engineering
- the emergent fields of the ontology and epistemology of technology and design
- engineering ethics and the ethics of specific technologies ranging from nuclear technologies to artificial intelligence
- written from philosophical and practitioners' perspectives and authored by philosophers and practitioners

The series also welcomes proposals that bring these fields together or advance philosophy of engineering and technology in other integrative ways.
Proposals should include:

- A short synopsis of the work or the introduction chapter
- The proposed Table of Contents
- The CV of the lead author(s) or editor(s)
- If available: one sample chapter

We aim to make a first decision within 1 month of submission. In case of a positive first decision the work will be provisionally contracted; the final decision about publication will depend upon the result of anonymous peer review of the complete manuscript. We aim to have the complete work peer-reviewed within 3 months of submission.
The series discourages the submission of manuscripts that contain reprints of previous published material and/or manuscripts that are below 150 pages/75,000 words.
For inquiries and submission of proposals authors can contact the editor-in-chief Pieter Vermaas via: p.e.vermaas@tudelft.nl, or contact one of the associate editors.

Levi Checketts • Benedict S. B. Chan
Editors

Social and Ethical Considerations of AI in East Asia and Beyond

Springer

Editors
Levi Checketts
Hong Kong Baptist University
Kowloon, Hong Kong

Benedict S. B. Chan
Department of Religion and Philosophy
Centre for Applied Ethics
Hong Kong Baptist University
Kowloon, Hong Kong

Editor-in-Chief
Pieter E. Vermaas

ISSN 1879-7202 ISSN 1879-7210 (electronic)
Philosophy of Engineering and Technology
ISBN 978-3-031-77856-8 ISBN 978-3-031-77857-5 (eBook)
https://doi.org/10.1007/978-3-031-77857-5

This Springer imprint is published by the registered company Springer Nature Switzerland AG
The registered company address is: Gewerbestrasse 11, 6330 Cham, Switzerland

Introduction: The Significance of East-West Dialogue on Artificial Intelligence

Abstract Recent trends in international research on Artificial Intelligence (AI) highlight the need for international ethical evaluations of AI. This chapter introduces the work of this volume—an intercultural dialogue on AI from Eastern and Western perspectives, emerging from an international conference in Hong Kong in 2023. The contributions in this volume represent genuine intercultural and international collaboration, and, as such, pose rich opportunities for policy makers, industry leaders, and academics alike to advance comparative ethical reflections. Perspectives include Confucian, Daoist, Buddhist, Western and Eastern Christian, Aristotelian and (post-)liberal positions, and focuses on AI ethics documents, large language models, sexbots, art, ghost workers, and existential threats. The following chapters can be read as individual contributions, though we believe they are better read as a broader conversation.

Keywords East-West dialogue • Artificial Intelligence • Hong Kong • Comparative ethics

After AlphaGo's successful defeat of world weiqi champion Lee Sedol in 2016, China and South Korea dedicated massive amounts of research funding to AI. Since then, China has surpassed the United States and Europe in percentage of AI journal publications cited globally. On a hopeful note for global dialogue, China-US collaborations on AI have increased fivefold over the past decade, and the two highest instances of current international collaboration in AI are between the United States and China and between the UK and China.

The global confluence of AI interest presents a corresponding global need for dialogue on ethical concerns for AI. In recent years, concerns regarding algorithmic bias, resource consumption, over-confidence in AI systems, and over-surveillance as immediate issues have overtaken older predictions about AI uprising, AI rights, or programming ethical systems. But while much applied AI research has become international, most international ethical AI work is still written through Western ethical perspectives.

Integrating East Asian philosophical traditions such as Confucianism and Daoism can significantly enhance our understanding of AI ethics. Confucianism emphasizes social harmony, human virtues, and ethical relationships. By incorporating virtues like *ren* (仁) and *li* (禮), Confucian ethics can address AI bias, transparency, and accountability, promoting fairness and the common good. Similarly, Daoism's focus on naturalness and balance offers a unique lens for AI ethics. Daoist ethics encourages the consideration of the harmonious relationship between humans and nature, urging sustainable AI practices and raising concerns about the impact of AI on the environment and human-nature connections.

These examples, while not exhaustive, demonstrate the valuable contributions of Chinese philosophy. Additional East Asian schools of thought can be used in the debates on AI, including Buddhism, Shinto, and even Asian Christianity. Moreover, comparisons between different Eastern and Western moral traditions, such as Confucianism and virtue ethics, can advance key work in AI and technology ethics. This approach allows for the fostering of a culturally diverse framework that promotes responsible AI development and deployment, acknowledging and respecting plural and diverse cultural perspectives. By doing so, we can better address the complexities of ethical challenges associated with AI.

From 30 March to 1 April 2023, the Centre for Applied Ethics held the first international conference at Hong Kong Baptist University (HKBU) after three full years of COVID-19 restrictions. Given the global interest in Artificial Intelligence at present, the organizers brought in scholars from across Asia, Europe, and the United States to talk about social and ethical concerns related to AI. Philosophers and theologians discussed AI concerns from within their tradition and social location, and the perspectives of sociologists and journalists provided necessary empirical descriptions of AI concerns they cover in their work. The impact of this conference was not negligible: for months following, many of the scholars and attendees cited the discussions from this conference in papers, other conferences, and collaborative projects.

This volume is the first fruit of this conversation. The reader should engage with these texts recognizing the rich tapestry of different perspectives here. In our initial preparations, we noted that the conversation surrounding ethical concerns of AI was varied and rich, but lacked much attention to perspectives from Asia. However, important as strictly Asian-perspective contributions are to this work, we saw a need for collaboration—a real dialogue between Western and Eastern thinkers, a confluence of religious perspectives, philosophical traditions and cultural backgrounds. In our opinion, we were able to reach this goal. Our conference featured, among the participants, a Black American theologian, a pioneering American philosopher of technology, a Japanese sociologist, a Filipino political philosopher, a Confucian bioethicist, an American-Hong Kong tech journalist, a Thai Buddhist AI ethicist, and many others. What is rare about such an eclectic array of thinkers, many of the participants engaged with each other over the course of the conference (and afterward) in genuine dialogue, taking interest in the work and thought of each other and finding inspiration for their own research.

Hong Kong, and specifically Hong Kong Baptist University, is the right place for this sort of project. Hong Kong has long stood as the gateway to China. Its colonial past has made it a vibrant city with a hybrid British-Chinese culture, a place where one can go across the harbor and see Chinese junk boats pull up to Colonial ferry piers, a place where the smell of incense could be coming from a street shrine or from an Anglican cathedral, a place where bright neon signs on twenty-story buildings overlook bustling night markets. After the 1997 handover to China, Hong Kong has remained a pivotal place for international business and political intercourse. And in recent years, with the interest of the mainland in AI development, Hong Kong has become a central place for science and technology innovation.

In reflecting on this project, we note that 1956 was an important year. It was the year Hong Kong Baptist College (changed in 1994 to be a public university) was founded, an institution dedicated to Christian whole person education. The initial mission was "to provide educational opportunities for the children of Hong Kong, especially those who have been unable to attend higher education due to economic situation, and to prepare them to be productive citizens of Hong Kong." At the same time, half the world away in Dartmouth, John McCarthy, Marvin Minsky, Nathaniel Rochester, and Claude Shannon were carrying out a summer workshop that is generally accepted as the birthplace of the field of AI. Depending on how one considers this year, then, one can either say HKBU is a "young" university or AI is an "old" technology, or neither or both. But this synchronicity offers a point of reflection: the convergence of minds, the dreams of the future, and the heritage we carry with us.

A few days before our conference, the Future of Life Institute released an open letter, calling for a cessation of all AI developments stronger than GPT4. The letter was signed by many luminaries, including Stuart Russell, Elon Musk, Andrew Yang, Max Tegmark, Steve Wozniak, and others. This call set the atmosphere for the conversation, but the work of scholars represented in these pages does not align with the jeremiad issued by Future of Life Institute (and promptly ignored by even their own signatories). Rather, the reader will find here a more reflective, honest approach to AI. AI neither signals the end of the world nor the coming utopia; it sits in a complex place between social and economic interests with some clear benefits and dangers, though more remains unclear. As one speaker at the conference offered, the language we use to refer to ethical issues in AI affects the way we address these issues. The reader will find in these pages language that attempts to highlight various concerns or questions rather than one-dimensional perspectives.

The diversity of perspectives presented here also reveals that the very existential challenge that AI poses—the question of the meaning of human life, the meaning of ethics, the meaning of intimacy, the future of human life, the function of language—are all points that are not so easily steamrolled into algorithmic expression. What does this mean? In spite of both optimists and pessimists, AI remains an open question. To call it the new god or to say that computers are people, as some of the conversations surrounding AI are wont to do, is to get it wrong. The critiques raised here all challenge the streamlined vision of humanity that often is compacted into discussions of AI. Thus, the answer to the challenge of AI offers an opportunity, as demonstrated here, to reflect on the complicated ways we think of ourselves, our world, and our use of technologies.

What we are left with then, is a challenge, a call, and a hope. The challenge is the call to not reduce the richness and profundity of human experience in the world—including our experiences of non-humans such as God, nature, history, and even our technologies. To borrow a concept from Catholic philosopher and *immortale* Jean-Luc Marion, we must recognize the "saturated phenomena" in the world that defy our linguistic frames (2004). The call then is to consider AI's place in a rich and complex account of human goodness. Any suggestion that AI should replace an ineffable aspect of human experience is to be resisted for rather a responsible adaptation of AI to human action. The hope is demonstrated in these pages and in the work of all who critically respond to the effort to frame AI as *the* question of humanity and not merely an opportunity for us to express more deeply that humanity.

This book is organized in four sections, which themselves reflect the international, intercultural project of this book. Part I opens the work with Chinese cultural perspectives. Pak Hang Wong begins his chapter by examining AI ethics policies and principles. Wong's concern, however, is directed toward the ambiguous goal of "trustworthy AI," an accomplishment everyone seems to agree is important but lack consensus on the meaning of this aim. Wong considers some various approaches but notes how most discussions favor Western assumptions of trust. Thus, he contends that a Confucian consideration of *xin* (信) provides an interesting nuance to the conversation by emphasizing the requirement that human trustees establish veracity in their behavior before they can be deemed trust-worthy. Such an insight shifts the consideration of "trustworthy AI" off of AI as trustworthy to purveyors of AI as needing to be trustworthy and avoid hype or "snake oil" claims.

Benedict S. B. Chan explores the ethical challenges of balancing privacy and information collection in the context of AI and technology ethics. He focuses on evaluating the contributions and limitations of Confucianism to the ongoing privacy debates, specifically addressing the issues of uncertainty and risk associated with privacy violations. By analyzing Confucianism's emphasis on intimate relationships, particularly within the family structure, Chan argues that Confucianism offers a unique perspective for addressing privacy concerns. However, it is also important to acknowledge the limitations of applying Confucianism to the privacy debates in AI and technology ethics, and Chan cautions against oversimplifying the discussion into an East-West dichotomy. Drawing from this analysis, his chapter explores practical examples, such as the ethical implications of Chatbots interacting with deceased individuals and the use of technology in education.

Finishing the section on Chinese thought, Levi Checketts addresses the hype surrounding Large Language Models and claims that they are "conscious" from Lacanian and Daoist perspectives. Using the models of the psyche from Jacques Lacan, Checketts addresses the fact that the processes of generating text in LLMs are indicative of a type of "thinking" characteristic of psychotics, those who do not live within the "symbolic order" typical of most human communication. Rather, Checketts argues that our understanding of AI as producing language is due to our own submission to the symbolic order. Drawing from the Daoist concept of *Jixin*

(機心), a word which means something like "machine-heart" while evading straightforward translation, Checketts shows that Daoist philosophy offers an alternative model of thinking beyond either the symbolic thought of most human language or the psychotic model of LLMs.

Brian Green opens the second section on Christian perspectives with a paper reflecting on the Pontifical Academy for Life's "Rome Call to AI Ethics" and the response from tech firms to implement their own AI ethics policies. Like Wong, Green considers how tech firms approach the question of AI ethics. Green is particularly well-situated to reflect on this, as his position as Technology Ethics Director at the Markkula Center for Applied Ethics requires him to work directly with stakeholders in the Silicon Valley community. Green finds that the basics of many corporations' ethics principles are similar to the Rome Call, but the broader Catholic ethical tradition has much more to say about AI ethics and both Church and companies would benefit tremendously from intentional efforts to dialogue, of which he suggests several useful approaches.

Robert James Boyles, a Filipino philosopher, contextualizes the international aspiration of the Rome Call within the specific context of the Philippines. While Filipino government, business and academic leaders have high expectations for the future of the island nation in the global AI community, Boyles argues that the country is not yet in a position to adhere to the general vision of the Rome Call. The record of the Philippine government on protecting human rights, promoting high-quality education, and maintaining transparency bodes poorly for its ability to realize the Rome Call. Boyles thus argues the nation will need massive reforms in these areas to minimally meet the expectations, but he is optimistic that the state may be beginning to move in that direction.

New Zealand theologian Stephen Garner likewise engages with the Rome Call but does so in a more ecumenical form through public theology. The concerns both Protestant and Catholic theologians have raised about the advent and use of AI should not be relegated merely to seminary conversations. Rather, public theology has as its aim engagement between the Church and society. Drawing a parallel with the bioethics tradition, Garner focuses on how theological concerns should influence policy decisions. Thus, concerns about human dignity, the plight of the poor, and abuse of power which theological ethicists have variously raised about AI should also inform both corporate development strategies and governmental regulation.

Much popular discourse surrounding AI addresses the question of its personhood. Pan-Chiu Lai, a Hong Kong theologian, addresses this question, and a concordant question of human control by or through AI by turning to Chinese Christianity as a source of understanding. Lai considers Confucianism and Christian theology to offer complementary views of the human person as relational, moral, spiritual, and transcendent. Or, in the word of Paul Tillich, the human person is multi-dimensional. Lai offers the nuanced view that this position takes into consideration the most advanced understandings we have from psychology, sociology, neuroscience, and other disciplines alongside the lasting wisdom from Confucianism (and Neo-Confucianism) and Christian theology. In the end, Lai rejects the reductive view of personhood attendant to AI conversations that would treat us as mere mathematical constructs and champions a positive view of interrelation cooperation and co-construction.

The third section engages in explicit comparative ethical questions surrounding AI. Opening this section is Shanghai-based philosopher Tianen Wang, who considers the revolutionary nature of Artificial Intelligence as upending all traditional forms of ethics. Most ethical approaches he contends are "world-adapting ethics," meaning we respond to the world as it is. With AI, human beings create new agents and new worlds, and so need a "world-creating ethic." Such an ethic would need to recognize the preexisting stipulations that do constrain us to avoid naïve overestimates of our capacity. It will also need to be far more universal than before, applied as a broad social ethics in a holistic manner—there can be no room for individual moralities as we create artificial beings.

Carl Mitcham asks the provocative question of whether liberalism is sufficient to reign in the problems of AI, a question that anticipates a forthcoming collection in this series dedicated to his contributions to philosophy of technology. Mitcham notes the threat of "existential risk" highlighted by many scientists, philosophers, and others and suggests that AI should be considered a technology in need of careful reigning in. However, liberalism, the leading political theory underwriting AI research, prioritizes individual freedom over collaborative, collective protections. As such, Mitcham worries that as long as liberalism is the reigning political ideology, existential threats, such as those posed by AI, are unlikely to be addressed, and alternative regimes are necessary.

Felix Yeung and Fei Song consider the seemingly intractable problem of machine ethics—what do we program "into" an AI to make it behave morally? They examine many of the debates that vex machine ethicists, no less trenchant than the same moral debates among ethicists broadly. But they note that the tendency toward "rationalist" ethical models favors a specifically Western (post-Enlightenment) trend and is not obviously the best solution. Rather, they propose virtue ethics as a paradigm, following the insights of Stuart and Hubert Dreyfus on skillful navigation in the world and what this means for programming machines. They make a compelling case that virtue allows us to not focus on rigid rules but rather the question of context and goal of the machines, avoiding problems such as moral hypocrisy and internal disagreement.

Soraj Hongladarom, similar to Pak Hang Wong, examines and interrogates the tendencies of AI ethics documents from a non-Western perspective. Engaging in cross-cultural studies, he finds that nearly all documents of principles outline "privacy, explainability, and lack of bias" as important goals of AI ethics. Noting that ethical principles do not appear out of thin air, Hongladarom engages in theoretical inquiry, asking the under-considered question of where these principles arise from and whether they are truly universal. He notes that some of these principles, such as privacy, have a particularly Western heritage, but that on the whole, they arise from the material existence of AI and the globalized problems that emerge from its construction. Thus, these issues really tie into other moral concerns which *are* universal, such as fairness and the good life. In the end, Hongladarom challenges AI ethics scholars to engage in cross-cultural inquiry to ensure that the principles they promote can be considered truly universal and on what basis.

This work concludes with a section considering the specific social or ethical concerns of the impact of AI. Karl Marx famously argued that "the point" of philosophy "is to change" the world. Often we philosophers get too caught up in theory and logical construction and lose focus on the situation in the real world. Hong Kong-based freelance journalist Karen Hao offers a very important corrective to the "ivory tower" view many of us gravitate toward. She has done investigative reports on "ghost workers" in poor countries who make AI "work" for us in rich countries. In particular, she interviewed and met with workers in Colombia and Kenya, and she reports on the fraught nature of contracted ghost work, the trauma induced by content moderation, the inegalitarian working conditions and general exploitation involved in the production of the shiny veneer that AI advocates tout. But Hao is optimistic that a better way of producing AI is possible and she offers suggestions such as unionization and more humanitarian emphasis on the part of contracting companies. At the time this work was started, Hao was a journalist for the *Wall Street Journal* and bound by contract to not publish elsewhere. The interview with her was conducted based on her presentation in the 2023 conference, a presentation that had reverberations in later research projects from many of the participants, including some in this volume. Such empirical research is critical for AI researchers from all cultural backgrounds to ground their theoretical critiques against real-world problems.

Art and aesthetics play an important role in our cultures and our understanding of ourselves. Japanese sociologist Takeshi Kimura interrogates what it means to say that artificial intelligence programs are now capable of producing outputs that can be aesthetically enjoyed. Kimura notes that previous discussions about the imposition of new technologies on art (e.g., Walter Benjamin's discussion of the camera) could not anticipate the emergence of AI art, where "prompt engineers" with no inherent artistic bodily skill are capable of generating works. Whether these can be considered "art" is difficult to say, but auction houses and art contests have thus far recognized such works as authentic. While Kimura does not offer a simple conclusion about this new era of aesthetics, he does delineate key issues that need to be addressed between the artistic and engineering communities.

The computer scientist David Levy speculates that, by 2050, people will have more sex with robots than with other human beings. Hong Kong philosopher Kai Man Kwan reflects on Levy's work in the final chapter, first exploring Levy and other "digi-sexuals" before considering the critiques made by feminists and traditional sexual ethicists about the expectation of a digi-sexual future. Kwan finds Levy's vision to be narrow, shallow on the question of relationality, dehumanizing and, ultimately, sexist. He cautions that Levy's vision of the future is not one that many people—from very different walks of life and perspectives—will find fulfilling.

Chinese philosophy of technology broadly and philosophy of AI specifically has seen significant development in recent years, mostly from the roots of Confucianism. Scholars like Wang (2021), Hung (2021), Li (2013), and others have built strong theoretical groundings of Confucian philosophy of technology. Yuk Hui (2016) engages in more direct intercultural constructive philosophy to build a new modern Chinese approach to the question. Few (Teschner & Tomasi, 2009) have considered adequately the ways Daoism contributes to this conversation, though Buddhism (Hershock, 2021) has more sway. This work, focused as it is on AI and not general attitudes toward technology, then contributes in a way to the work done by Wong (2021), Hongladarom (2020), Zhu et al. (2020), and others, but does so with a specific interest in cultivating East-West dialogue. While the present volume does not attempt to offer a deep theoretical framework for carrying out this work, it does provide facets of this conversation as they take place; insights from people working in intersections of cultures and traditions, people of good faith willing to engage in dialogue that assumes they come from a cultural background without assuming this is the only background that exists.

While many readers may opt to select individual chapters for their own merit, a narrative can be found tying together the different pieces that make up this volume, not in spite of but because of the various perspectives that abound. Artificial Intelligence is a technology whose advent has polarized groups of supporters and doom-sayers, a phenomenon that Science and Technology Studies (STS) scholar Lee Vinsel (2021) refers to as "Criti-Hype." Against the promises of everything new or the warnings of impending destruction we offer nuanced considerations of how to think about AI. And against a universalizing tendency to treat all problems the same, we offer a pluralist account, one where all-too-human moral worries rebuff the instrumentalist rationality regnant of one-size-fits-all algorithmic thinking. Instead of wrestling with whose voice is most important, we offer a variety of voices. To an outsider, this may seem just another romantic East-West dialogue, but to those of us based in Hong Kong, this is the very essence of our culture. Thus, we offer this narrative as a glimpse into what AI ethics from a Hong Kong perspective looks like.

Hong Kong Baptist University
Kowloon, Hong Kong SAR

Levi Checketts
Benedict S. B. Chan

References

Future of Life Institute. (2023). *Pause giant AI experiments: An open letter*. https://futureoflife.org/open-letter/pause-giant-ai-experiments/

Hongladarom, S. (2020). *The ethics of AI and robotics: A Buddhist viewpoint*. Lexington Books.

Hershock, P. (2021). *Buddhism and intelligent technology: Toward a more humane future*. Bloomsbury Academic.

Hui, Y. *The question concerning technology in China: An essay in cosmotechnics*. MIT Press.

Hung, C. Technological mediation *in* and *for* Confucianism-based cultures. In P-H. Wong, & T. X. Wang (Eds.), *Harmonious technology: A Confucian ethics of technology* (pp. 50–64). Routledge.

Li, C. (2013). Confucian perspectives on science and technology. *Encyclopedia.com* https://www.encyclopedia.com/science/encyclopedias-almanacs-transcripts-and-maps/confucian-perspectives

Marion, J-L. (2004). *In excess: Studies of saturated phenomena* (R. Horner, & V. Berraud, Trans.). Fordham University Press.

Teschner, G., & Tomasi, A. Technological Paradigm in Ancient Taoism. *Techné: Research in Philosophy and Technology, 13*(3), 190–205.

Vinsel, L. (2021). You're doing it wrong: Notes on criticism and technology hype. *STS News*. https://sts-news.medium.com/youre-doing-it-wrong-notes-on-criticism-and-technology-hype-18b08b4307e5

Wang, T. X. (2021). Confucian ritual technicity and philosophy of technology. In P-H. Wong, & T. X. Wang (Eds.), *Harmonious technology: A Confucian ethics of technology* (pp. 10–28). Routledge.

Wong, P-H. (2021). Artificial intelligence, personal decisions, consent, and the Confucian idea of oneness. In P-H. Wong, & T. X. Wang (Eds.), *Harmonious Technology: A Confucian Ethics of Technology* (pp. 79–94). Routledge.

Zhu, Q., et al. (2020). Blame-laden moral rebukes and the morally competent robot: A Confucian ethical perspective. *Science and Engineering Ethics, 26*, 2511–2526.

Acknowledgements

The editors acknowledge and express their gratitude for the support of the Centre for Applied Ethics (CAE), the Department of Religion and Philosophy (REL), the Master of Arts in Ethics and Public Affairs (MAEPA) programme, and the former Faculty of Arts (now Faculty of Arts and Social Sciences) at Hong Kong Baptist University. Institutional support has been crucial for the organization of the conference, gathering the articles, proofreading, and communicating with our authors. In addition to those contributors included in the volume, our thanks go to Mary Ho Yan Siu, Saiming Wong, Andrew Wai Luen Kwok, Kam Ho Chan, Ascar Ng and others whose work and support were crucial for both a successful conference and collecting together the papers of this volume. We also thank Pieter Vermaas, series editor-in-chief and other editors and staff at Springer for working with us so well in this process. This work, beginning from planning stages in late 2022 to completion at the start of 2025 was only possible with an entire network of supporters.

Contents

Part I
Chinese Philosophical Approaches to AI

Chapter 1
Confucian 'Trustworthy AI': Diversifying a Keyword in the Ethics of AI and Governance

Pak-Hang Wong

Abstract The term 'trustworthy AI' has become one of the keywords in the academic and policy discussion of the social, ethical, and legal issues of artificial intelligence (AI), but it remains a contested notion and there are open debates about its meaning and how it is to be operationalized. For philosophers and ethicists, the debate has particularly revolved around the meaning of 'trust' and 'trustworthiness'. However, since a *thin* and *acultural* understanding of trust and trustworthiness has been common in the debate, the notion of trustworthy AI is at best ambiguous in a *global* and *cross-cultural* context, where trust and trustworthiness can be interpreted differently. A culturally-sensitive exploration of trust, trustworthiness, and trustworthy AI is, therefore, much needed to gain insights on *what and how trustworthy AI can be conceptualised through the normative resources of other cultural traditions*. This article attempts to offer a culturally sensitive exploration of trustworthy AI by drawing on recent analysis of trust and trustworthiness in Confucian philosophy. More specifically, I shall demonstrate how taking seriously the Confucian view of trust and trustworthiness in the context of AI provides us an alternative way to think about the ethics and governance of AI

Keywords Trustworthiness · Confucian Ethics · *Xin* · Trustworthy AI · AI Governance

---

P.-H. Wong (✉)
Department of Religion and Philosophy, Centre for Applied Ethics,
Hong Kong Baptist University, Kowloon, Hong Kong
e-mail: pakhangwong@hkbu.edu.hk

L. Checketts, B. S. B. Chan (eds.), *Social and Ethical Considerations of AI in East Asia and Beyond*, Philosophy of Engineering and Technology 47,
https://doi.org/10.1007/978-3-031-77857-5_1

1.1 Introduction

The term 'trustworthy AI' has become one of the keywords in the academic and policy discussion of the social, ethical, and legal issues of artificial intelligence (AI). As various AI ethics principles and guidelines, as well as AI policies and regulations, propose and/or prescribe (human) values to be considered or (social and technical) requirements to be fulfilled in the design, development, and deployment of AI systems for them to be *trustworthy*, trustworthy AI has arguably become synonymous to the objective of these guidelines, principles, policies, and regulations (see, e.g., Schmitt, 2022). While trustworthy AI has now been well-entrenched in the academic and policy discussion of AI, it remains a contested notion and there are open debates about its meaning and how it is to be operationalized (Mökander & Floridi, 2021; Braun et al., 2021; Thiebes et al., 2021; Kaur et al., 2022; Liu et al., 2022; Laux et al., 2023; Díaz-Rodríguez et al., 2023).

For philosophers and ethicists, the discussions have particularly revolved around the meaning of 'trust' and 'trustworthiness' in the context of AI. Interestingly, the discussion seldom acknowledges that the meaning of trust and trustworthiness may mean something different for different communities and cultures (see, e.g., Ferrin & Gillespie, 2010; Kwantes & Kuo, 2021), and it often assumes a *thin* and *acultural* understanding of trust and trustworthiness. In this respect, the notion of trustworthy AI not only remains ambiguous in a *global* and *cross-cultural* context insofar as trust and trustworthiness can be interpreted differently by different communities and cultures, but its applicability can also be called into question when the cultural subtleties about trust and trustworthiness are unaccounted for. At the very least, a culturally sensitive exploration of trust, trustworthiness, and trustworthy AI shall offer insights on *what and how trustworthy AI can be conceptualised through the normative resources of other cultural traditions*, thereby extending the current discussion on the topic.

This article attempts to offer one such culturally sensitive exploration of trustworthy AI by drawing on recent analysis of trust and trustworthiness in Confucian philosophy. More specifically, I shall demonstrate how taking seriously the Confucian view of trust and trustworthiness in the context of AI provides us an alternative way to think about the ethics and governance of AI. I proceed by reviewing the prominence of 'trustworthy AI' in recent AI policy and regulation around the world, and then briefly summarize the existing philosophical and ethical discussion on trustworthy AI, especially in relation to the notions of trust and trustworthiness. Next, I introduce a Confucian view on trust and trustworthiness, and then end the article by applying it to construct an account of Confucian trustworthy AI.

1.2 "Trustworthy AI" in the Ethics and Governance of AI

Long before the term 'trustworthy AI' was coined and used, the social, ethical, and legal implications of AI had already been extensively explored by researchers from various academic disciplines (see, e.g., Brey & Søraker, 2009; Müller, 2020; Stahl, 2022). Indeed, terms such as "ethical AI", "AI for good", and "beneficial AI", and, most notably, "responsible AI" have already been used in the writings about the ethics and governance of AI before 'trustworthy AI' came to prominence.[1] 'Trustworthy AI' differs from the other terms by emphasizing the role of trust as a basic condition for people to adopt the technology, thereby realizing the opportunities of its applications. As the EU High-Level Expert Group on AI (AI HLEG) stated, "human beings and communities will only be able to have confidence in the technology's development and its applications when a clear and comprehensive framework for achieving its trustworthiness is in place" (AI HLEG, 2019, 4).

The *Ethics Guidelines for Trustworthy AI* (Ethics Guidelines) published by the AI HLEG offered the first detailed analysis of the meaning of trustworthy AI and made the term popular. In the Ethics Guidelines, trustworthy AI is described as consisting of three components, namely being *lawful*, *ethical*, and *technically and socially robust*. Lawful AI is AI applications that are in compliance with applicable laws and regulations, ethical AI is AI applications which adhere to ethical principles and the European Union's values, and finally, robust AI refers to AI applications that perform in a safe, secure, and reliable manner with appropriate measures and safeguards to foresee and prevent unintended adverse impacts of the AI applications (AI HLEG, 2019, 6–7). The Ethics Guidelines focused on the component of ethical AI, and elaborated further on the ethical principles and European Union's values in the context of AI, including the principle of respect for human autonomy, the principle of prevention of harm, the principle of fairness, and the principle of explicability, as well as various rights set out in the Charter of Fundamental Rights of the European Union (AI HLEG, 2019, 9–13).

Beyond the foundation of trustworthy AI, the Ethics Guidelines also elaborated on seven *substantive* requirement for AI applications to meet for them to be considered as trustworthy, namely: (1) human agency and oversight, which requires AI applications to have appropriate level(s) of human control; (2) technical robustness and safety, which requires AI to be resilient, secure, and reliable; (3) privacy and data governance, which requires AI applications to respect privacy and data protection and accompanied by data governance mechanisms; (4) transparency, which requires AI applications to be built and implemented in a transparent and explainable manner; (5) diversity, non-discrimination and fairness, which requires AI applications to avoid unwarranted biases and involve stakeholders in their building and implementation; (6) societal and environmental well-being, which requires AI applications to account for their *long-term* social and environmental impacts; and, finally, (7) accountability, which requires mechanisms to be in place to ensure

[1] See, Stix (2022) for an account of the development of these terminologies.

responsibility and accountability for AI applications and their outcomes.[2] Here, the substantive requirements proposed by the Ethics Guidelines captured specific *ethical* and *technical* concerns of AI applications as well as *what* the problem of trust and trustworthiness is in the context of AI and *how* the problem is to be addressed. In this respect, the seven substantive requirements outline a *thicker* meaning of trustworthy AI that ties to the purpose of building and implementation of AI applications.

The term 'trustworthy AI' has been used in various policy documents and regulations around the globe since the publication of the Ethics Guidelines.[3] For instance, trustworthy AI is presented as the objective of the *OECD AI Principles*. The OECD AI Principles aim to "promote [the use of AI] that is *trustworthy* and respects human-centred and democratic values" (OECD, 2019). Similarly, the Chinese *White Paper on Trustworthy Artificial Intelligence* [可信人工智能白皮书], co-authored by China Academy of Information and Communications Technology and JD Explore Academy, described a crisis of trust triggered by AI and the importance of trustworthy AI as a response to the crisis of trust (CAICT & JD Explore Academy, 2021). The *Artificial Intelligence Risk Management Framework*, published by the National Institute of Standards and Technology (NIST) of the United States Department of Commerce, is also intended to be "a resource to the organizations designing, developing, deploying, or using AI systems to help manage the many risks of AI and promote trustworthy and responsible development and use of AI systems" (NIST, 2023, 2).

Interestingly, the values proposed by these documents share many similarities with the seven substantive requirements in the Ethics Guidelines. For example, the OECD AI Principles mentions (1) inclusive growth, sustainable development and well-being; (2) human-centred values and fairness; (3) transparency and explainability; (4) robustness, security and safety; and (5) accountability. The Chinese *White Paper on Trustworthy AI* includes (1) reliability & controllability; (2) transparency & explainability; (3) data protection; (4) accountability; and (5) diversity & inclusiveness. Finally, the *AI Risk Management Framework* characterizes trustworthy AI as being (1) valid and reliable; (2) safe; (3) secure and resilient; (4) accountable and transparent; (5) explainable and interpretable; (6) privacy-enhanced; and (7) fair with harmful biases managed. Certainly, there are clear differences between these documents in how they construe and individuate various values in relation to trust, trustworthiness, and trustworthy AI, but it is also clear that there is a

[2] For a detailed exposition of the requirements, see AI HLEG (2019), 14–20. In addition to the requirements, the Ethics Guidelines also discuss various methods to operationalize trustworthy AI and assess AI applications' trustworthiness. As I am concerned primarily with the philosophical and ethical discussion of trust and trustworthiness in relation to trustworthy AI, the methods to operationalize and/or assess trustworthy AI are not within the scope of the current article.

[3] In terms of AI policy and regulation, the *Proposal for a Regulation of the European Parliament and of the Council Laying Down Harmonised Rules on Artificial Intelligence (Artificial Intelligence Act) and Amending Certain Union Legislative Acts*, commonly known as the EU AI Act, presents the most straightforward application of trustworthy AI. In fact, the EU AI Act is drafted on the basis on the Ethics Guidelines, and thus the notion of trustworthy AI figures prominently in the Act.

convergence in the values they have included as contributing to AI being trustworthy. Here, the convergence suggests there is a shared view on *what* is the problem of trustworthy AI and *how* to address it; in particular, the successful inclusion of these values in *the building and implementation* of AI applications is regarded as the key to achieving trustworthy AI. While the shared view of trustworthy AI is useful to foster (global) collaboration in building, implementing, and governing AI, it may also foreclose alternative views of trust and trustworthiness in the context of AI insofar as the alternative views fall outside the purview of the shared view. So construed, it is important to also explore alternative views of trust and trustworthiness in the context of AI to allow for a more comprehensive discussion of trustworthy AI for the ethics and governance of AI.

1.3 Philosophical Discussion of Trust and Trustworthiness in "Trustworthy AI"

Although a shared view on trustworthy AI and its importance and relevance for building, implementing, and governing AI has started to emerge in policy and regulatory landscape as well as in the popular and academic discourse, the notion of trustworthy AI and the related ideas of trust and trustworthiness remain a subject of debate. In particular, the leading AI policy documents and regulations and AI ethics principles and guidelines have provided substantive requirements and values for AI applications to be deemed as trustworthy, but few offer a more detailed account of trust and/or trustworthiness which connects them to the proposed substantive requirements and values.[4] In this respect, the "objective list frameworks" (Simion & Kelp, 2023) or "formulaic approach" (Braun et al., 2021) to trustworthy AI, which primarily aim to enumerate the requirements and values of trustworthy AI, do not sufficiently answer why and how the requirements and values enable people to trust the AI applications or why and how exactly they contribute to their trustworthiness.

One of the key questions, therefore, is the *nature* and *meaning* of trust and trustworthiness in the context of AI, and this question has been discussed by researchers from different backgrounds. Here, my focus is on the *philosophical and ethical* debate surrounding the nature and meaning of trust and trustworthiness in the context of AI. Since there are already several excellent critical reviews of the debate (see, e.g., Reinhardt, 2023; Freiman, 2022), I shall only highlight a few *key* questions in the debate among philosophers and ethicists:

I. Whether trust in AI is possible and/or plausible? If it is indeed possible (and/or plausible), what does 'trust' mean in the context of AI?

[4] However, there have been discussions on the ideas of trust and trustworthiness entailed by the major frameworks, such as EU's *Ethics Guidelines for Trustworthy AI* (see, e.g., Rieder et al., 2021) and NIST's *Artificial Intelligence Risk Management Framework* (see, e.g., Newman, 2023).

II. Whether trustworthiness can be attributed to AI—a non-human entity? And, what features are required AI to be capable of being trustworthy?
III. What is, if any, the relation between trust and trustworthiness in the context of AI? In particular, does trustworthiness of AI entail trust in AI and *vice versa*?

Wildly different answers have been given to (I), (II), and (III). For example, Joanna Bryson (2018) argues that trust is a relation between peers, and insofar as people and AI are not peers, people *cannot* trust AI (also, see Ryan, 2020). Relatedly, Rieder et al. (2021) argue that being trustworthy has both *moral* and *epistemic* components, in which an agent (or entity) can be considered as trustworthy only when it is trust-responsive and truthful as well as knowledgeable and free from epistemic vices. Since AI cannot (yet) exhibit some of these qualities, it cannot be deemed as trustworthy. They do, however, consider the broader socio-technical arrangement of an AI application as a plausible candidate for being trustworthy. Ramón Alvarado, on the other hand, suggests that epistemic trust—a specific kind of trust, which "is allocated strictly on the basis of the capacities of the receiver as a provider of information or as a conveyer of knowledge" (Alvarado, 2022, 1)—is the only kind of trust that *can* be used in properly characterizing the trust relationship between human and AI. Regarding the relationship between trust and trustworthiness in the context of AI, Karoline Reinhardt (2023) notices that the current debate is one-sided and overly emphasizes the role of trustee, as the focus has been on how trust is established by trustees being trustworthy. Reinhardt also argues that trust-building is a two way, reciprocal exercise that the role of both trustee and trustor is equally significant. Margit Sutrop makes a similar point when she describes trusting as "a communication process.... On the trustee's side it involves expression of the trustee's competence and good will, as well as her/his commitment to do what is expected. On the trustor's side, trust involves a desire that the trustee will prove to be trustworthy, a positive feeling towards the trustee, and a participant stance to act in a certain way (either feeling betrayed or gratified)" (2019, 508).

However (I), (II), and (III) are to be settled, it is surprising that few in the discussion have paid attention to the *cultural* dimension in thinking about trust and trustworthiness, i.e., that they can be interpreted differently in different cultural traditions. The missing cultural dimension in the philosophical and ethical debate on trust and trustworthiness in the context of AI can, in part, be attributed to the search for an account of trust and trustworthiness that is applicable globally without needing to account for the potential differences and divergence in the meaning of these ideas. Doing so, unfortunately, also leads to missing opportunities to learn from other cultural traditions in their understandings of trust and trustworthiness, especially how different understandings of trust and trustworthiness may help us to (re)interpret and (re)formulate trustworthy AI.

1.4 A Confucian View of "Trustworthiness"

In Confucian philosophy, *xin* [信] is arguably the term with the meaning closest to the idea of 'trustworthiness' in English.[5] *Xin* has a key role in Confucian philosophy: it is one of the "four teachings" of Confucius and also one of the "five constants": benevolence (*ren*, 仁), righteousness (*yi*, 义), ritual propriety (*li*, 礼), wisdom (*zhi*, 智) and trustworthiness (*xin*, 信).[6] Indeed, *Analects* 2.22 is often mentioned as supporting the importance of *xin* in Confucianism,

> The Master said, "I do not know how a man without [trustworthiness] is to get on. How can a large carriage be made to go without the crossbar for yoking the oxen to, or a small carriage without the arrangement for yoking the horses?" (Legge, 1861)

Despite *xin*'s importance in Confucianism, there are only a handful of philosophical analysis of *xin*-as-trustworthiness in comparison to other Confucian virtues such as *ren* and *yi*. We can, nonetheless, find two recent insightful attempts to analyze *xin* in Cecilia Wee (2011) and Winnie Sung (2020).

Wee (2011) analyses the meaning of *xin* as the term appears and is being used in the *Analects*, and argues that *xin* is tied to abiding by one's speech. From this understanding of *xin*, together with her elaboration of Confucius' view of the importance of appropriate speech and how speech reflects a person's character, Wee proposes an account of Confucian *xin* person as someone who "is evidently careful when making a verbal undertaking. She assures that she is able to fulfill an undertaking before making one, and does not promise beyond what she can perform" (2011, 523). In addition, she points out that a Confucian *xin* person's doing as she says she will should not to be driven by its consequence but by an intrinsic concern of abiding by one's speech, which is what distinguishes *xin* from reliability or being reliable.

Pace Wee, Sung (2020) argues that the scope of *xin* extends well beyond abiding by one's speech.[7] Instead, Sung argues what is characteristic of Confucian *xin* is in effect a "matching requirement" that requires "a match between how one [consciously] presents oneself and the way one in fact is" (2020, 274). In this respect, Sung's account of *xin* shares with Wee's (2011) in its emphasis on *xin* as a self-directed concept, i.e., *xin* is concerned with one's matching self-presentation (or, in Wee's account, abiding by one's speech) regardless of the consequence of doing so, and, indeed, regardless of whether there is a specific person who trusted the person of *xin*. However, Sung also argues that insofar as there is a potential audience of one's self-presentation, *xin* also entails an active (epistemic) respect for others,

---

[5] *Xin* has also been interpreted as "faithfulness", "good faith", "being true to one's word", "fidelity" or "truthfulness" (Lu, 2001) and "reliability" (Lai, 2018). This article does not aim to argue for a specific translation or interpretation of the term, but to build on existing accounts of *xin*-as-trustworthiness, and discuss its implications to the discussion surrounding trustworthy AI.

[6] For a brief historical overview of *xin*, see Wang et al. (2020).

[7] A similar critique to the narrow understanding of *xin* as abiding by one's speech is offered by Lu (2001). It is argued that the narrow understanding does not corroborate with the fact that *xin* is one of the "five constants" in Confucianism, where its scope ought to be broader than one's speech.

because the person of *xin* would be sensitive to the limits of how one can come to know about another person. As self-presentation is one of the main source of information for others to know about her, being *xin* is in fact about providing others reliable information about who one is and *enabling* them to decide how they want to relate to her.

While Wee (2011) and Sung (2020) differ in their analysis of *xin*, they have both foregrounded an *active* dimension of being a trustee—or, more accurately, of being a person of *xin*. In Wee's account, a person of *xin* ought to attend to what she says and promises and not to overstate or over-promise. In Sung's account, a person of *xin* ought to properly, in the sense of matching, present oneself to others. What is common in both cases is that they are *active* tasks of a person of *xin*. Moreover, it is also worth noting that in both the account of Wee and Sung, being *xin* precedes one's uptake of the role of trustee, and so is conceptually distinct from trust. It is the active dimension of *xin* as well as its conceptual distinctiveness, I think, that can offer interesting insights to the ethics and governance of AI, especially to our thinking of trustworthy AI.

1.5 From Confucian Trustworthiness to Confucian Trustworthy AI

The conceptual distinctiveness of *xin-as-trustworthiness* and its *active* dimension offer theoretical, normative and practical insights into the philosophical and ethical discussion of trustworthy AI. Before expounding the insights, however, I shall briefly return to the questions (I), (II), and (III) highlighted previously with the concept of *xin*. Doing so can help situate the Confucian view within the broader discussion of trustworthy AI and set the ground for expounding the insights. For (I), the Confucian view of *xin* is without doubt about human beings, and thus is inapplicable to AI *per se*, but, I contend that the Confucian view remains applicable and relevant to the discussion of trustworthy AI to the extent that we follow Rieder et al. (2021) in thinking about trustworthiness of trustworthy AI as something for the socio-technical arrangement of an AI application but not of the application itself—with the human beings being the bearers of trustworthiness. The Confucian view is also pertinent to (II); in particular, as Sung (2020) has proposed, the capability of active (epistemic) respect for others is necessary for *xin*, and this capability has yet to receive attention in the discussion of trust and trustworthiness in the context of AI. Finally, the Confucian view seems to suggest there is *no* straightforward entailment from *xin-as-trustworthiness* to trust. It does, however, emphasize the role of people-as-trustors in the forming of trust relationship by deferring to them and their judgment on whether to form and/or enter into a trust relation.

I shall now expound in more detail the theoretical, normative, and practical insights into the discussion of trustworthy AI from Confucian view of *xin*. *Xin*'s being prior to a trust relationship suggests *xin* would play a more fundamental normative role than it is proposed in the Ethics Guidelines (and other similar policies, regulations, ethics guidelines, and principles) that trustworthiness of AI applications aim to allow people to have confidence in them. In contrast, *xin-as-trustworthiness* entails a responsibility of critical self-scrutiny in speech (Wee, 2011) or in self-presentation (Sung, 2020) with *or without* people *actually* being trustors; and, in the context of AI, it requires AI applications to critically scrutinize their promises and images such that the chance of overstatement and misrepresentation is minimized. To reiterate, the aim of *xin-as-trustworthiness* is to empower people to decide whether to enter into a trust relation, i.e., being a trustor of someone, or not.

The responsibility of critical self-scrutiny, together with the active dimension of *xin-as-trustworthiness*, in turn highlight a particular issue in the ethics and governance of AI that has not been adequately addressed, namely the problem of *AI hype* and *snake oil*. An AI hype can be understood as an overly inflated expectations about the capabilities of AI (or of a specific AI application) that distracts from the precise way in which AI works and can genuinely contribute to people and society (Siegel, 2023), whereas AI "snake oil" refers to AI applications that "[do] not and cannot work" (Narayanan, 2019). AI hype and snake oil are harmful, as they encourage propagation of non-functional and/or dysfunctional AI applications that underperform and even are counterproductive through false promises and presentations of unrealistic or misleading capabilities. For example, a general harm can ensure a replacement of a functional (manual) practice with a non-functional (or, worse, dysfunctional) AI application *because of* the hyped capabilities.[8] Moreover, more often than not, the harm will be borne by those who are marginalized and/or belong to vulnerable group(s), as it is well known that the cost and burden of AI are unevenly distributed (see, e.g., Kaltheuner, 2021). In addition, AI hype and snake oil may divert attention from the more reality-grounded questions to more speculative ones (see, e.g., Krafft, 2020).

AI hype and snake oil can be viewed as a failure to critical self-scrutiny both of the talk(s) and image(s) of the providers of AI applications—or, of tech companies in general. In this respect, the Confucian view presented in this article does offer normative requirements to providers of AI applications: they ought to verify and validate the capabilities of AI before they *talk* (or advertise) about its capabilities, and they ought to ensure the users of their AI applications as well as the public to have an appropriate *image* of their AI applications. More importantly, as Sung (2020) has proposed, being *xin-as-trustworthiness* requires an active (epistemic) respect for others; and, in the context of AI, the active (epistemic) respect ought to be given *not only* to the potential users of an AI application, who will be the ones to decide whether or not to enter into a trust relation with the socio-technical arrangement of the AI application; active (epistemic) respect also needs to be given to those

[8] Raji et al. (2022) discuss a related issue to the ethics and governance of AI in terms of presumed functionality of AI applications. They have noted that very few researchers have explicitly acknowledged AI applications' being non-functional and dysfunctional is equally important as a topic for the ethics and governance of AI as other topics such as fairness, explainability, etc.

who will be affected—either directly or indirectly—by the AI application, which, in turn, allows them to decide on their relations with the socio-technical arrangement of the AI application, e.g., entrusting, resistance.

The emphasis on talking about and presenting *in a genuine way* the capabilities of AI applications also highlight the need to *communicate* and *showcase* trustworthiness of AI applications. In other words, the Confucian view calls for a rethinking of how trustworthiness of an AI application can be appropriately communicated and showcased to its users and the public (see, e.g., Liao & Sundar, 2022)

In short, *xin-as-trustworthiness* offers an alternative normative foundation of trustworthy AI *vis-a-vis* critical self-scrutiny, a path to the underexplored issue of AI hype and snake oil in the ethics and governance of AI, as well as a specific need for trustworthy AI to communicate and showcase trustworthiness of AI applications.

1.6 Conclusion

This article is set to offer a culturally sensitive exploration of trustworthy AI through the lens of Confucian philosophy. I have described a Confucian view of *xin-as-trustworthiness*, and its potential implications for rethinking the notion of trustworthy AI.I In particular, I suggest that the Confucian view will assume a different *objective* of trustworthy AI, i.e., *xin-as-trustworthiness* is significant not only for people to be(come) confident in AI applications, but to empower them in their judgments on forming and entering into trust relations with the providers of AI applications.

Before ending this article, I should note that it is *not* an attempt to offer a comparative analysis of trust and/or trustworthiness from a Confucian perspective to the "European perspective" (or the "American perspective") and argue for the superiority of one over the others. The objective of the current article, to reiterate, is to draw insights from recent research on Confucian view of *xin-as-trustworthiness*, and apply them in the context of AI. This exercise, I contend, is a useful exercise to demonstrate how looking at the normative resources of other cultural traditions can enrich the discussion of the ethics and governance of AI, and thereby calling for similar exercise for other normative notions and in other cultural traditions.

References

AI HLEG (High-Level Expert Group on Artificial Intelligence). (2019). *Ethics guidelines for trustworthy AI*. Available at: https://digital-strategy.ec.europa.eu/en/library/ethics-guidelines-trustworthy-ai

Alvarado, R. (2022). *What kind of trust does AI deserve, if any?* AI and Ethics.

Braun, M., Bleher, H., & Hummel, P. (2021). A leap of faith: Is there a formula for "Trustworthy" AI? *The Hastings Center Report, 51*(3), 17–22.

Brey, P., & Søraker, J. H. (2009). Philosophy of computing and information technology. In D. M. Gabbay, P. Thagard, J. Woods, & A. W. M. Meijers (Eds.), *Philosophy of technology and engineering sciences* (pp. 1341–1407). Elsevier.

Bryson, J. (2018). *AI & global governance: No one should trust AI*. UNU-CPR. Available at: https://unu.edu/cpr/blog-post/ai-global-governance-no-one-should-trust-ai

CAICT (China Academy of Information and Communications Technology) & JD Explore Academy. (2021). *White paper on trustworthy artificial intelligence*. Available at: http://www.caict.ac.cn/english/research/whitepapers/202110/t20211014_391097.html

Díaz-Rodríguez, N., Del Ser, J., Coeckelbergh, M., López de Prado, M., Herrera-Viedma, E., & Herrera, F. (2023). Connecting the dots in trustworthy artificial intelligence: From AI principles, ethics, and key requirements to responsible AI systems and regulation. *Information Fusion, 99*, 101896.

Ferrin, D., & Gillespie, N. (2010). Trust differences across national–societal cultures: Much to do, or much ado about nothing? In M. Saunders, D. Skinner, G. Dietz, N. Gillespie, & R. Lewicki (Eds.), *Organizational trust: A cultural perspective* (pp. 42–86). Cambridge University Press.

Freiman, O. (2022). *Making sense of the conceptual nonsense 'trustworthy AI'*. AI and Ethics.

Kaltheuner, F., Birhane, A., Raji, I. D., Amironesei, R., Denton, E., Hanna, A., Nicole, H., Smart, A., Oduro, S. D., Vincent, J., Reben, A., Milne, G., Black, C., Harvey, A., Strait, A., Parida, T., Ashok, A., Jansen, F., Cath, C., & Peppin, A. (2021). *Fake AI*. Meatspace Press.

Kaur, D., Uslu, S., Rittichier, K. J., & Durresi, A. (2022). Trustworthy artificial intelligence: A review. *ACM Computer Survey, 55*(2), 39.

Krafft, P. M., Young, M., Katell, M., Huang, K., & Bugingo, G. (2020). Defining AI in policy versus practice. In *AIES '20: Proceedings of the AAAI/ACM Conference on AI, Ethics, and Society* (pp 72–78).

Kwantes, C. T., & Kuo, B. C. H. (2021). *Trust and trustworthiness across cultures*. Springer.

Lai, K. (2018). Learning to be reliable: Confucius' Analects'. In K. Lai, R. Benitez, & H. J. Kim (Eds.), *Cultivating a good life in early Chinese and Ancient Greek philosophy: Perspectives and reverberations* (pp. 193–207). Bloomsbury Publishing.

Laux, J., Wachter, S., & Mittelstadt, B. (2023). *Trustworthy artificial intelligence and the European Union AI act: On the conflation of trustworthiness and acceptability of risk*. Regulation & Governance.

Legg, J. (1861). *The Analects*. Available at: https://ctext.org/analects

Liao, Q. V., & Sundar, S. S. (2022) Designing for responsible trust in AI systems: A communication perspective. In *FAccT '22: Proceedings of the 2022 ACM Conference on Fairness, Accountability, and Transparency* (pp. 1257–1268).

Liu, H., Wang, Y., Fan, W., Liu, X., Li, Y., Jain, S., Liu, Y., Jain, A., & Tang, J. (2022). Trustworthy AI: A computational perspective. *ACM Transactions on Intelligent Systems and Technology, 14*(1), 59.

Lu, Z. (2001). Fiduciary society and confucian theory of Xin—On Tu Wei-ming's fiduciarity proposal. *Asian Philosophy, 11*(2), 85–101.

Mökander, J., & Floridi, L. (2021). Ethics-based auditing to develop trustworthy AI. *Minds & Machines, 31*, 323–327.

Müller, V. C. (2020). *Ethics of artificial intelligence and robotics*. The Stanford Encyclopedia of Philosophy. Available at: https://plato.stanford.edu/archives/fall2023/entries/ethics-ai/

Narayanan, A. (2019). *How to recognize AI snake oil*. Arthur Miller Lecture on Science and Ethics.

Newman, J. (2023). A taxonomy of trustworthiness for artificial intelligence. *CLTC White Paper Series*. Available at: https://cltc.berkeley.edu/publication/a-taxonomy-of-trustworthiness-for-artificial-intelligence/

NIST. (2023). *Artificial Intelligence Risk Management Framework* (AI RMF 1.0). Available at: https://airc.nist.gov/AI_RMF_Knowledge_Base/AI_RMF

OECD. (2019). *Recommendation of the council on artificial intelligence*. Available at: https://legalinstruments.oecd.org/en/instruments/OECD-LEGAL-0449

Raji, I. D., Kumar, I. E., Horowitz, A., & Selbst, A. (2022).The fallacy of AI functionality. In *FAccT '22: Proceedings of the 2022 ACM Conference on Fairness, Accountability, and Transparency* (pp. 959–972).
Reinhardt, K. (2023). Trust and trustworthiness in AI ethics. *AI and Ethics, 3*, 735–744.
Rieder, G., Simon, J., & Wong, P.-H. (2021). Mapping the stony road toward trustworthy ai: Expectations, problems, conundrums. In M. Pelillo & T. Scantamburlo (Eds.), *Machines we trust: Perspectives on dependable AI* (pp. 27–40). The MIT Press.
Ryan, M. (2020). In AI we trust: Ethics, artificial intelligence, and reliability. *Science and Engineering Ethics, 26*(5), 2749–2767.
Schmitt, L. (2022). Mapping global AI governance: A nascent regime in a fragmented landscape. *AI and Ethics, 2*, 303–314.
Seigel, E. (2023). The AI hype cycle is distracting companies. *Harvard Business Review*. Available at: https://hbr.org/2023/06/the-ai-hype-cycle-is-distracting-companies
Simion, M., & Kelp, C. (2023). Trustworthy artificial intelligence. *Asian Journal of Philosophy, 2*, 8.
Stahl, B. C. (2022). From computer ethics and the ethics of AI towards an ethics of digital ecosystems. *AI and Ethics, 2*, 65–77.
Stix, C. (2022). Artificial intelligence by any other name: A brief history of the conceptualization of 'Trustworthy Artificial Intelligence.'. *Discover Artificial Intelligence, 2*, 26.
Sung, W. (2020). Xin: Being trustworthy. *International Philosophical Quarterly, 60*(3), 271–286.
Sutrop, M. (2019). Should we trust artificial intelligence? *Trames, 23*(4), 499–522.
Thiebes, S., Lins, S., & Sunyaev, A. (2021). Trustworthy artificial intelligence. *Electronic Markets, 31*, 447–464.
Wang, Y., Bao, Q., & Guan, G. (2020). Trustworthiness (xin, 信). In *History of chinese philosophy through its key terms* (pp. 117–124). Springer.
Wee, C. (2011). *"Xin"*, trust, and confucius' ethics. *Philosophy East and West, 61*(3), 516–533.

Chapter 2
A Preliminary Investigation of Confucianism's Impact on Privacy Debates in A.I. and Technology Ethics

Benedict S. B. Chan

Abstract With the widespread availability of personal information on the internet and its potential unauthorized disclosure, the rapid advancement of technology has brought forth numerous ethical challenges. Balancing the conflict between privacy and information collection is a pressing concern that necessitates thoughtful consideration from all stakeholders. Philosophers play a crucial role in providing a moral framework to address this conflict. While ongoing debates on privacy persist in the Western context, this chapter focuses on evaluating the contribution and limitations of Confucianism to these discussions. Specifically, it examines the debate's central issues of uncertainty and risk associated with privacy violations and highlights the significance of the intimacy account within Confucian thought. By analyzing Confucianism's emphasis on intimate relationships, particularly within the family structure, this chapter argues that it offers a unique perspective for addressing privacy concerns. However, it also acknowledges the limitations of applying Confucianism to the privacy debates in A.I. and technology ethics, cautioning against oversimplifying the discussion into an East-West dichotomy. Drawing from the aforementioned analysis, this chapter further explores the evaluation of privacy issues in practical examples and applications within the realm of A.I. and technology ethics. These examples include the ethical implications of Chatbots interacting with deceased individuals and the utilization of technologies in education in both physical and virtual classrooms.

Keywords Confucianism · Intimacy · Privacy · Chatbots of the Dead · Technologies in Education

B. S. B. Chan (✉)
Department of Religion and Philosophy, Centre for Applied Ethics, Hong Kong Baptist University, Kowloon, Hong Kong
e-mail: benedictchan@hkbu.edu.hk

L. Checketts, B. S. B. Chan (eds.), *Social and Ethical Considerations of AI in East Asia and Beyond*, Philosophy of Engineering and Technology 47,
https://doi.org/10.1007/978-3-031-77857-5_2

2.1 Introduction

Nowadays, our information may easily be found on the internet and revealed to others without our consent. Alongside greater convenience, the rapid development of technology in the modern world has also brought about many ethical problems. Contact Tracing Apps, i-beacon technology, A.I. robots, and ChatGPT are just a few examples of technologies that can be employed by governments or organizations to collect and exploit individuals' information for various purposes. Defining privacy is not a straightforward task, as there is no single definition of privacy (Francis & Francis, 2017, 1–29). Privacy encompasses various aspects, including physical or body privacy and information privacy. A descriptive approach to privacy views it as a family resemblance, where privacy is defined by the different aspects it encompasses. On the other hand, the normative approach to privacy highlights its importance as a moral value. How to handle and balance the conflict between privacy and collecting information (or even surveillance) is a topic that everyone should think about. And it is important for philosophers to provide a moral framework to solve such a conflict.

While the debates are still ongoing in the West, the objective of this essay is to illustrate how specific elements of Confucianism can contribute to the global and local development of privacy, as well as examine their limitations. In Chinese contexts, "privacy" is an expression combining two different words: "szu" (私) and "yin" (隱). Szu refers to the private realm of individuals and families, while yin describes the act of hiding or concealing one's actions or thoughts (Whitman, 1985). While linguistic analyses may provide a starting point for understanding privacy in Confucianism, it is not entirely useful to assume that this is precisely how Confucians conceptualize privacy. Indeed, it is debatable whether we can separate Chinese words and discuss each of them in Confucian texts in such a manner and assume that they pertain to privacy. The issue here is that the classical Confucian texts do not have an exact equivalent term for "privacy" in Chinese. A useful analogy is the United States Constitution, which also does not explicitly mention the term "privacy" in its text. As a result, legal disputes have arisen as to whether privacy is a constitutional right. The key difference is that the United States Supreme Court has interpreted certain provisions of the Constitution as implying a right to privacy (e.g., The First, Fourth, and Fifth Amendments), while Confucianism lacks such an authority to reach a similar consensus.

Rather than merely providing a historical and linguistic analysis, this chapter focuses on the philosophical analysis behind Confucianism and privacy. It discusses the ongoing debate surrounding the problems of uncertainty and risk in violating privacy and highlights the contribution of the intimacy account to this debate. By examining Confucianism's emphasis on intimacy relationships, particularly in terms of family values, the chapter argues that it can provide a unique perspective on addressing the problems, which can be called a Confucian intimacy account.

Nonetheless, the chapter also acknowledges the limitations of applying not only Confucianism but also other East Asian traditions in the debate on privacy, and argues against the simplistic East-West dichotomy in discussing privacy. I also apply the ideas developed in this chapter to discuss privacy issues in practical examples within the ethics of A.I. and technology. These examples include using chatbots of the deceased for commemorating the departed and the utilization of technologies in education, such as teaching in physical and virtual classrooms and even the merits and drawbacks of having an imagined A.I. robot teacher.

The structure of this essay is outlined as follows: Sect. 2.2 will address the common (mis-)understanding of the East-West dichotomy in discussing privacy, which typically says that while the West emphasizes privacy from an individualistic or autonomous standpoint, the East places greater emphasis on relationships and community. The limitations of such a dichotomy will be discussed. Instead of this simple dichotomy, I propose delving deeper into the discussion of the intimacy component of privacy, and subsequently exploring how Confucianism can provide insights in this regard. Moving on to Sect. 2.3, the essay will exlopre the challenges of uncertainty and risk in relation to privacy violations. The contribution of the intimacy account to this debate will be explained. Section 2.4 will focus on what we can call the Confucian intimacy account. I will discuss the role of intimacy relationships within Confucianism, and how Confucian principles can contribute to the discourse surrounding privacy violations discussed in Sect. 2.3. In Sect. 2.5, the essay will apply the aforementioned discussions to practical scenarios such as the use of chatbots of the dead and A.I. and technologies in education, elucidating how Confucianism can offer contributions to these specific cases.

2.2 Against Oversimplifying the East-West Dichotomy on Privacy

In this section, I aim to examine the common misunderstanding of the East-West dichotomy regarding privacy. Indeed, this dichotomy extends beyond privacy to encompass moral traditions and considerations in both the East and the West. Since the focus of this chapter is on privacy, I will primarily confine the dichotomy to privacy debates. Generally, it is often depicted that the West prioritizes privacy as individual freedom and autonomy from surveillance, whereas the East is perceived as placing more emphasis on relationships and community. In short, the East-West dichotomy revolves around the debate between the so-called "individualism" and "collectivism." While this dichotomy may hold relevance in certain contexts and situations, it risks oversimplifying the complexities of Eastern and Western moral traditions. Therefore, it is imperative to critically analyze this prevailing perspective and take into account other influential factors, rather than reducing it to a mere cultural distinction. Further exploration is necessary to delve more deeply into these points, laying the groundwork for the argument in later sections of this chapter.[1]

As mentioned above, the prevailing understanding of the East-West dichotomy often assumes that the West uniformly prioritizes individualism, emphasizing freedom and autonomy. In simple terms, Western privacy is viewed as the protection of personal matters, with the belief that others should not interfere with an individual's choices. For instance, when examining concepts like privacy and freedom, Yuanye Ma discusses their connection with individual autonomy, as defined in Kantian ethics (2019, pp. 198–201). Kantian ethics places a strong emphasis on individual autonomy and the moral duty to respect others as rational beings with their own agency. In relation to privacy, Kantian ethics would suggest that individuals have the right to control their personal information and make choices about what is shared with others. This notion of privacy aligns with some Western understandings of protecting personal matters and ensuring that others do not interfere with an individual's choices. If we limit our understanding of freedom to the Kantian framework, it may lead to the misconception that such ideals are exclusive to the West, while East Asian traditions lack a thorough analysis of these concepts.

However, it is crucial to recognize that the Western perspectives on privacy are not solely grounded in Kantian individualism. Within Western thoughts, a variety of interpretations of freedom and privacy exist that diverge from Kantian ethics. For example, while utilitarianism, as advocated by J.S. Mill, also values freedom, it approaches the concept differently from Kantian ethics (Mill, 1859). Utilitarianism emphasizes maximizing overall happiness or utility for the greatest number of people. In terms of privacy, this perspective may lead to a different prioritization. For instance, utilitarian reasoning might justify intrusions into individual privacy if it results in greater overall happiness. The debate over whether utilitarianism is inherently individualistic is ongoing, as it is often contended that utilitarianism allows for more instrumental harm than Kantian deontology (Kahane et al., 2018; Mihailov, 2023). This distinction showcases how the utilitarian view of freedom and privacy can deviate from the Kantian framework, highlighting the diversity of perspectives within Western traditions.

In certain Western moral frameworks, such as feminist caring ethics, privacy is valued as a means to nurture communal relationships and foster a sense of belonging (Noddings, 2013; Gilligan, 1982). By highlighting these points, we can also challenge the oversimplified notion that the West exclusively associates privacy with individualism. It is worth noting that East Asian traditions may approach the concepts of freedom and privacy differently from Kantian ethics or utilitarianism. A later discussion will compare feminist caring ethics with Confucianism, offering insights into the diverse moral traditions and philosophical theories involved, which will provide a more nuanced understanding beyond a strict East-West dichotomy.

[1] Some references can be considered as using the context of the East-West dichotomy (Hongladarom, 2016; Lü, 2005; Ma, 2019; Wang, 2019; Whitman, 1985). However, some of these authors provide additional elaborations on this dichotomy, and their arguments are valuable for further discussion. I will explore some of these arguments in detail in later sections.

A more significant concern with the prevailing understanding of the East-West dichotomy is the portrayal of the East as primarily prioritizing relationships and community, neglecting individual privacy. While some East Asian traditions do emphasize the interconnectedness of individuals and the importance of communal harmony, it does not imply a disregard for individual privacy completely. If we consider the concept of freedom at face value, without exclusively adopting Kantian or utilitarian perspectives, it becomes debatable whether East Asian traditions completely disregard the value of freedom. Scholars are engaged in ongoing debates regarding whether Confucianism encompasses certain concepts of freedom. Christina Whitman, for example, argues that Confucianism does not prioritize freedom from surveillance and control (1985, pp. 92–97). However, Chenyang Li (2014) draws upon concepts of freedom from the feminist caring approach, compares them with Confucianism, and posits that Confucian notions of freedom involve the freedom to choose the good (*ze shan* 擇善). Robert A. Carleo III also suggests that we can consider Confucianism as endorsing freedom as the "subjective affirmation of the good" and "the cultivation of desires aligned with the good" (Carleo, 2021, p. 211). These perspectives demonstrate that the question of whether East Asian traditions value freedom is not as straightforward as it may initially appear. And even if it is really the case that there is an absence of explicit discussions on privacy as freedom from surveillance in East Asian traditions, this does not signify a complete disregard for individual freedom and privacy.

The points discussed above collectively serve as illustrative examples that support a common direction, highlights the importance of avoiding an oversimplified and exaggerated East-West dichotomy when examining privacy. While it is true that privacy may be understood and valued differently within the context of maintaining harmonious relationships and social balance in the East, it is imperative to conduct a thorough analysis of the concepts of privacy and East Asian traditions. In doing so, we should narrow the scope of the debate and focus on the arguments at hand. The subsequent sections of this chapter will undertake this task by first focusing on the problems and solutions concerning the violation of privacy. Subsequently, we will explore the contribution and limitations of Confucianism to the privacy discourse. Finally, we will apply these findings to examine specific cases within the realm of A.I. and technology ethics.

2.3 Violation of Privacy and the Intimacy Account

To begin with, let me briefly summarize some ideas on the meanings of the violation of privacy. There are two main accounts used to explain the meaning of the violation of privacy: the control account and the access account (Allen, 2011; Inness, 1992; Solove, 2008; Macnish, 2018). The control account emphasizes that privacy entails control over one's personal information. It means that if one loses control over their information, they have lost a degree of privacy. On the other hand, the access account holds that for there to be an actual violation of privacy, information needs to be accessed without authorization or consent. This account emphasizes that privacy is fundamentally about protection from intrusion and information gathering by others. It is worth noting that the moral issues involved in privacy violations often involve both concepts of control and access. For instance, during the COVID-19 pandemic, concerns have arisen regarding how personal data will be accessed by schools or society, as well as concerns related to the data being controlled by others (Chan, 2022). While the access account appears to provide a better explanation of what constitutes a privacy infringement, the control account also captures people's concerns and anxieties. The ongoing debate between the control account and the access account has yet to reach a definitive conclusion. Recent works have proposed further views that attempt to reconcile the two accounts (Mainz & Uhrenfeldt, 2022; Lundgren, 2020; Menges, 2021; Munch & Mainz, 2023).

Regardless of whether it pertains to the control account or the access account, the loss of control or access over personal information presents two main problems. First, it can lead to feelings of vulnerability and uncertainty, which can undermine an individual's sense of security. Second, there are issues of risk, which suggest that even in the absence of concrete evidence that personal information has been controlled or accessed, individuals may assume that their privacy has been violated. These two problems highlight the common concerns and anxieties that underlie the importance of individuals having control or access over their personal information. Generally, the concerns here also highlight the complex nature of privacy and the need for ongoing discussions and considerations in ethics of A.I. and technology (Macnish, 2018, pp. 424–429).

Given that uncertainty (or insecurity) and risk are two common concerns related to privacy violations, it is imperative to identify and address these issues. Generally, there are three key elements that warrant consideration when addressing privacy concerns, each addressing different aspects. The first element revolves around trust, the second pertains to consent, and the third focuses on justice. For the purpose of this chapter, the exclusive focus will be on the first element, namely trust.[2] In discussing trust, it is necessary to discuss the concept of the intimacy account of privacy (Gerstein, 1978; Inness, 1992; Rachels, 1975; Reiman, 1976). The remainder of this section will examine the role of the intimacy account in the privacy debate, followed by the subsequent section exploring the relationship between the intimacy account and Confucianism.

Numerous scholars have investigated the relationship between intimacy and privacy. For example, Robert Gerstein (1978) contends that intimate relationships would be unattainable if privacy were absent. Judith Thomson (1975) acknowledges how intimacy makes privacy a unique and indispensable value, despite her position that privacy is not a moral right. James Rachels (1975) argues that privacy empowers individuals to determine which information about them is disclosed, facilitating the establishment of diverse interpersonal relationships, including intimate ones.

[2] For the discussions of the second and third elements, see (Zimmer, 2010; Mills, 2022; Nussbaum, 2003).

Jeffrey Reiman (1976) rebuts Rachels' argument but still affirms the significance of interpersonal relationships, particularly intimate ones, to privacy. Generally, privacy is a complex concept that involves various dimensions, including freedom from surveillance, as well as the protection of intimate relationships. The concept of intimacy is a significant aspect of privacy that is frequently discussed in philosophical debates. Intimacy pertains not only to romantic and marital relationships but also to other close relationships. The intimacy account of privacy emphasizes the importance of safeguarding the intimate aspects of individuals' lives, such as their personal relationships, feelings, and thoughts. Without privacy, it would be difficult to establish and maintain meaningful intimate relationships. The relationship between privacy and intimacy is reciprocal: as people become more intimate, they are more willing to give up their privacy between each other, and the information shared between them can be more freely exchanged. However, for non-intimate relationships, we tend to value privacy more, and sometimes even use the degree of privacy reserved to judge how close or distant our relationship is. In short, the interplay between privacy and intimacy is intricate: while privacy is a prerequisite for intimate relationships, intimacy simultaneously transforms privacy into an exclusive and indispensable value.

In addition, it is crucial to consider the relationship between intimacy and the issues addressed by the control and access accounts. As mentioned above, while the control and access accounts shed light on the nature of privacy violations, the concept of intimacy can offer a valuable perspective in resolving these challenges. The access account emphasizes the significance of authorized access to personal information, while the control account highlights the importance of individuals maintaining control over their personal data. Both accounts have their merits and limitations, but they can be interconnected with the intimacy account. For example, Julie Inness (1992) subscribes to both the control account and the intimacy account, recognizing their complementary nature. Conversely, Kevin Macnish (2018) challenges Inness's control account while acknowledging the relevance of intimacy in privacy discussions. Macnish presents an illustrative example to highlight the distinctions between these accounts (Macnish 2018, p. 420). Consider a situation where an individual leaves her diary unattended on a coffee shop table for 30 minutes and returns to find a stranger holding her diary. The control account suggests that privacy has been compromised because the person lost control over her diary during that time, leaving it susceptible to various possibilities. On the other hand, the access account argues that privacy violation only occurs if the stranger actually reads (accesses) the diary without authorization. While the control account may appear more widely accepted and explored in existing literature, Macnish contends that the access account better explains privacy violations.

An important insight arises from this particular scenario is that the diary holds significance because it may contain personal and intimate information. Therefore, revealing such information to strangers would indeed constitute a violation of privacy. The reciprocal relationship between intimacy and privacy underscores the crucial role of trust in instances of privacy violation. For example, the content of a diary being exposed to unknown individuals, or to individuals known to the owner but with whom the owner has never shared the diary's information, can be a serious breach of privacy. Conversely, if the diary is exposed to a person whom the owner trusts and regularly shares the diary's information with, the issue may not be as severe. Trust plays a vital role as individuals rely on the assurance that their personal information will be handled responsibly and not used against them. In this context, the intimacy account aligns with both the control and access accounts by framing privacy within the framework of trust. It addresses the uncertainties and mitigates the risks associated with privacy violations. By considering the role of trust, the intimacy account offers a valuable perspective in comprehending and resolving the challenges posed by privacy violations.

In this section, I have briefly outlined the current debate on the violation of privacy. As mentioned, the intimacy account does not address all privacy concerns, such as issues of consent and justice. Nevertheless, the intimacy account recognizes that individuals encounter various uncertainty and risk, but intimate relationships often mitigate those uncertainty and risk of losing information. Therefore, intimacy is a part of the solution to solve the problems of uncertainty and risk. In next section, I will discuss how Confucianism contributes to such a discussion of the violation of privacy.

2.4 Confucianism and the Intimacy Account

In this section, I will explore the relationship between Confucianism and the intimacy account, which will lead to what I call the Confucian intimacy account. First, I will provide an overview of Confucian perspectives on intimacy and close relationships, such as those found within families. I will then continue the critique discussed in Sect. 2.2, which is about the common assumption that Western philosophy is centered around individualism, while Confucianism is focused on collectivism and relationships. Rather, I will argue that Confucianism contributes to the development of moral concepts that can be used to address ethical issues. Particularly, I will discuss how Confucianism can be applied to the intimacy account discussed in the previous section, as well as to issues related to privacy.

As previously discussed, privacy is a multifaceted concept with no single definition. The best way to understand the relationship between Confucianism and privacy is not through linguistic analyses, but rather through considering the ethical issues or problems related to privacy that we encounter, and then assessing how Confucianism serves as a valuable cultural and moral foundation for the intimacy account. Confucianism is well-known for its emphasis on defending and promoting family relationships, which are considered some of the strongest forms of intimate relationships. The concept of privacy is important for strengthening the bond within intimate relationships, including those within a family.

As Whitman correctly observes, privacy is valuable because "it allows people the freedom to define these relationships in ways that suit their own needs, relatively immune from social constraints" (Whitman, 1985, p. 89). When we consider the importance of close relationships, Confucianism provides a philosophical tradition that emphasizes their significance. Confucianism places great emphasis on the importance of family, with filial piety being a central concept. The classical text *Xiao Jing* (《孝經》) is dedicated to exploring the importance of filial piety, with its opening passage stating that "(It was filial piety.) Now filial piety is the root of (all) virtue, and (the stem) out of which grows (all moral) teaching."[3] Confucianism thus provides a framework for valuing and promoting strong family relationships, which are considered the foundation of a virtuous and harmonious society. As the first chapter of *Da Xue* (*The Great Learning* 大學) states, "Wishing to order well their states, they first regulated their families."[4] This view underscores the importance of cultivating virtuous family relationships as the foundation for a harmonious and well-ordered society. For example, in *Analects* 13:18, Confucius notes that the father should conceal the misconduct of his son, and vice versa, as this demonstrates uprightness. The Chinese word of conceal in this passage is "*yin*" (隱). For another example, in *Analects* 8:13, Confucius also says that "When right principles of government prevail in the kingdom, he will show himself; when they are prostrated, he will keep concealed."[5] In this passage, it does not treat the concealing as something negative, but rather a prudent way to act when the way (i.e., the *Dao*) does not prevail.

Additionally, Confucianism also provides a valuable moral foundation for defending and justifying the importance of intimate relationships. Confucianism traditionally endorses the "five relationships," which can be considered as an extension of family relationship. These five relationships are: father and son, sovereign and minister, husband and wife, older and younger brothers, and friends. Each relationship is related to some virtues. For example, in *Mencius* 3A4, it writes "between father and son, there should be affection; between sovereign and minister, righteousness; between husband and wife, attention to their separate functions; between old and young, a proper order; and between friends, fidelity."[6] In addition, Chap. 20 of *Zhong Yong (The Doctrine of Mean* 中庸*)* also says: "The duties are those between sovereign and minister, between father and son, between husband and wife, between elder brother and younger, and those belonging to the intercourse of friends. Those five are the duties of universal obligation."[7] As such, it is essential for proponents of an intimacy account to explore the moral principles underlying this perspective within Confucianism.

While Confucianism provides a valuable moral foundation for understanding the concept of intimacy, it is important to recognize that the intimacy account does not rely solely on Confucianism as the exclusive moral basis. Furthermore, it is not

---

[3] Unless otherwise specified, the translations of Chinese classical texts used in this section are sourced from James Legge's translation, which can be accessed through the "Chinese Text Project" available on the internet. See https://ctext.org/xiao-jing

[4] See https://ctext.org/liji/da-xue

[5] See https://ctext.org/analects/tai-bo

[6] See https://ctext.org/mengzi/teng-wen-gong-i

[7] See https://ctext.org/liji/zhong-yong

necessary to assume that the classical Confucian texts explicitly address privacy or the intimacy account; thus, this analysis is not solely a linguistic analysis of the Confucian texts. Moreover, it is crucial to acknowledge that Confucianism possesses its unique interpretations of the meaning of family, filial piety, and intimate relationships. Employing Confucianism as a framework for understanding intimacy does not negate its distinctiveness. For example, according to Whitman, this Confucian perspective on intimate relationships emphasizes the harmony and union within the relationships, and each person does not exist as a separate sphere. As she argues, Confucianism has overlapping rules governing both family life and public life. For each person, "Through his ties to his family, the individual, at least in theory, cannot draw totally apart, leaving the rest of the world behind" (Whitman, 1985, pp. 89–90). In short, one cannot separate individual from family, or separate family from the public life.

However, as discussed in Sect. 2.2, while it is undeniable that cultural distinctions exist between the East and the West, it is inappropriate to attribute these differences solely to the binary distinction between collective or relational and individualistic orientations. For instance, Confucianism also emphasizes the significance of the individual, as demonstrated in *Da Xue*, which also emphasizes the importance of self-cultivation.[8] On the other hand, certain Western moral theories are difficult to classify as purely individualistic, rendering the term "individualism" ambiguous and imprecise. Feminist caring approach, as an example, does not promote individualism as its core principle.

After comprehending these ideas, we can apply them to our discourse on Confucianism and privacy. While Whitman (1985) has provided a detailed explanation of how Confucianism endorses intimate relationships, her approach assumes that Western culture is individualistic and Confucianism is collective, which may oversimplify the comparison. To avoid such a simplistic comparison, we should move forward and consider alternative perspectives. Yuanye Ma (2019) also proposes what she calls a relational privacy framework, which provides a detailed explanation of the concepts of privacy in terms of relationship. She argues for the concept of "relational person," which can be based on either feminism or Confucianism. Her aim is to show that relational privacy, which is based on the concept of the relational person, is a better account of privacy than individual privacy. By emphasizing the importance of relationships in privacy, Ma challenges the traditional focus on individual autonomy in Kantian ethics.

Ma's approach offers some noteworthy points. First, she compares Kantian, feminist, and Confucian thoughts on privacy, which is a clearer and more nuanced approach than relying on simplistic comparisons between the West and the East. Second, Ma's research on Confucianism and relational privacy provides valuable

[8] As mentioned before, *Da Xue* writes in the first passage that "Wishing to regulate their families, they first cultivated their persons." While there are then four more virtual items to explain further how to do self-cultivation, it also says clearly that "all must consider the cultivation of the person the root of everything besides. It cannot be, when the root is neglected, that what should spring from it will be well ordered." https://ctext.org/liji/da-xue

insights for further research and inquiry on how Confucianism emphasizes the moral importance of various kinds of relationships (Ma, 2019, pp. 199–201). However, her approach still has some limitations. While Ma avoids using the unclear framework of the West and the East, she seems to assume that Confucianism only emphasizes on relationships and then uses that as the representation of East Asian traditions to deal with the concept of privacy. This assumption is still unclear because it reduces Confucianism to mere relational ethics, ignoring its complexity and diversity. The only part that may escape this problem in her work is the comparison between Confucianism and feminist caring approach, but this is also controversial. While some scholars advocate for this comparison, others believe that it is problematic (Li, 2022, 2002, 1994; Star, 2002; Yuan, 2002). Given that such a comparison complicates the discussion of privacy issues, we should be cautious about putting feminist caring approach and Confucianism on the same side in the debates on privacy.

Additionally, a more fundamental issue is that dividing privacy issues into individual and relational autonomy/person still leads to oversimplification. Focusing solely on relational privacy may overlook other important aspects of privacy. One of the primary challenges, as previously noted, is the uncertainty and risk involved in violating privacy, regardless of whether using control or access accounts. This creates many doubts and fears that must be addressed. Although discussing relationships can be helpful in resolving some issues, it alone has limitations. It is important to also explore other concepts such as consent and justice, as previously noted. Focusing solely on relational privacy without considering individual autonomy or other relevant privacy concepts may not fully address concerns about uncertainty and risk. This may even regress to a previous discussion of East vs. West, which cannot adequately resolve all privacy concerns. This can be considered as the limitation of relational approach to privacy, or even the intimacy account of privacy.

Therefore, while Ma's approach has some merits, we should be careful about the scope of the approach. What we can do is to first address the issues related to privacy (e.g., uncertainty and risk), and then explore some possible solutions (e.g., the intimacy account), then figure out how Confucianism can help to solve these problems (e.g., Ma's research on relational privacy is useful for this part). This is more reasonable than dividing privacy merely into individual or collective aspects.

2.5 Applications: Chatbots of the Dead and Technologies in Education

Now let us discuss two examples that apply the ideas mentioned above to practical situations involving A.I. and technology ethics. The first example concerns the ethics of using chatbots to simulate deceased individuals. Currently, there is a debate regarding whether it is morally acceptable to utilize artificial intelligence to recreate the presence of deceased loved ones. On one hand, supporters argue that it can offer closure and comfort to grieving individuals by allowing them to interact with a virtual representation of their loved ones, aiding in the grieving process and providing emotional support. Some may even view it as a potentially necessary medical technology for close relatives to recover from emotional traumas caused by the loss of deceased individuals. On the other hand, critics argue that creating artificial versions of deceased individuals without their consent is unethical and may exploit emotionally vulnerable individuals. Concerns include potential misuse and manipulation of these chatbots, as well as questions about the authenticity of the interactions and their impact on the grieving process. These concerns also raise issues related to privacy and consent, specifically regarding the collection of information about the deceased individuals and the communication between the deceased individuals and their close relatives or friends. Privacy concerns arise from the potential gathering of personal information about the deceased, which may include sensitive or private details. Additionally, the interactions between the deceased individuals and their loved ones may involve personal conversations and exchanges, raising questions about the consent of both the deceased individuals and their families in using such information in a simulated context. This debate serves as a practical application of our discussion in this chapter.

How does Confucianism contribute to this debate? Alexis Elder (2020) argues from a Confucian perspective that chatbots simulating the deceased should be morally acceptable in principle, as long as there are regulations and limitations in place to define their boundaries of use. Elder's argument can be summarized as follows. First, in comparison to other East Asian traditions like Daoism or Mohism, Confucianism places greater emphasis on grieving for the deceased who had close relationships with us. Additionally, while both Confucianism and Aristotelianism emphasize the importance of being virtuous persons, Confucianism specifically focuses on the dictates of etiquette and ritual. Putting these ideas together, Elder argues that Confucian regards grieving for important relatives as an essential feature of being a virtuous person, and hence "Much of the value of funeral and grieving rituals was to be found in their contributions to the development and sustenance of these features." (Elder, 2020, p. 78, see also pp. 76–80).

Second, Elder discusses the details of grieving rituals for the deceased drawing upon Xunzi's ideas. For instance, Xunzi argues that these rituals provide us with the necessary time to mourn and adapt. Despite the artificial nature of many aspects of these rituals, Confucianism considers them valuable. Elder argues that the mourning process discussed by Xunzi involves elements of "impersonation and scripted responses" on behalf of the deceased, which aids in facilitating the mourning process. In a similar vein, if chatbots can assume the role of digital impersonators for the deceased, they could enable grieving individuals "to express appreciation even after death, and model the expressing of thanks to the bereaved in turn, in a way that is (or can be) both sorrowful and respectful" (Elder, 2020, p. 82). In this sense, chatbots may serve a comparable purpose as the rituals Xunzi discussed.

Lastly, Confucian also endorses some limitations on the uses of the chatbots of the dead. For example, the purpose of producing chatbots of the dead should aim to "help people carry their valued pasts forward with them into the future" but not for photorealism or even the replace the deceased individuals (Elder, 2020, p. 86). With proper limitations and regulations, Confucian would agree that we should not worry the use of chatbots of the dead.

Elder's Confucian argument regarding the use of chatbots simulating the deceased presents an intriguing perspective. Unfortunately, due to space constraints in this chapter, I am unable to thoroughly explore the soundness of his argument. However, let us assume that his argument holds merit. What I find noteworthy is the parallel and resonating nature of his argument with the discussions on Confucianism and intimate relationships within this chapter. While Elder does not extensively address the privacy concerns associated with the utilization of chatbots simulating the deceased, it is reasonable to extend his argument to encompass the intersection of Confucianism and the intimacy account within the broader privacy debate.[9]

One can argue that privacy concerns in the context of chatbots simulating the deceased extend beyond the mere collection of personal information about the deceased individuals. They also encompass the interactions between the deceased individuals and their close relatives or friends. In this regard, questions of consent and boundaries become crucial considerations. However, it is worth exploring the relationship between intimacy and privacy when addressing these concerns. Drawing from Elder's argument, which permits the use of chatbots of the deceased for grieving and mourning purposes, we can apply a similar approach to the privacy issues surrounding these chatbots. In other words, we can consider treating the information and privacy concerns of the chatbots based on the intimate relationship between the deceased and their relatives, friends, or family members. Just as rituals are conducted to memorialize the deceased, the chatbots can contain information about the deceased, enabling people to mourn and adapt. By utilizing the framework of intimate relationship according to Confucianism, we can establish boundaries and guidelines for the use of chatbots simulating the deceased, ensuring that privacy is respected while allowing individuals to engage in the grieving process. It requires careful consideration of the dynamics within the intimate relationship and the consent of the deceased, as well as the preferences and needs of the living individuals involved. This approach acknowledges the significance of intimacy and privacy in the context of using chatbots to facilitate mourning and highlights the need for a nuanced and ethical exploration of these complex issues. The Confucian intimacy account discussed in this chapter can contribute to such an approach.

The second example I want to discuss in this chapter is about Confucianism and education in A.I. and technology ethics. Nowadays, technologies are increasingly intertwined with higher education. The use of information technologies, such as big data analytics, automated decision-making systems, and online teaching technologies, offers various opportunities to enhance the teaching and learning experience in both virtual and physical classrooms. However, these new technologies also challenge important ethical concepts, especially in terms of privacy, such as the privacy

[9] His discussion on privacy can be found in Elder (2020, pp. 74 & 85), although both references are relatively brief and not directly connected to Confucianism.

of students and teachers in their own spaces and the collection of sensitive data and big data. The situation has become more severe during the COVID-19 pandemic, as many people are confined to their homes to attend online classes. I have been collaborating with scholars from other disciplines on interdisciplinary research into the ethics of privacy when using technologies to facilitate engaging teaching and learning experiences in virtual and physical classrooms. If students are more familiar with their teachers, they tend to be more confident in using online platforms such as Zoom and classroom technologies such as iBeacon technology, and do not feel as though their privacy is at risk. Although this finding may be expected, it serves as an illustration of how a stronger, more intimate relationship between students and teachers can enhance trust and mitigate concerns related to uncertainty and risk (Lam et al., 2022; Chan 2020, 2022; Chan et al., 2021, 2022; Griffiths et al., 2019).

While the preceding research primarily focuses on technology ethics in education, it is also possible to extend this discussion to encompass artificial intelligence. One such extension is presented by Charlene Tan (2020), who examines whether from the Confucian perspective, A.I. can assist or potentially replace human teachers in spiritual education. Tan suggests using A.I. to create a "digital Confucius," an imagined robot teacher well-versed in Confucian texts such as the *Analects* and other classical works. To assess its effectiveness in comparison to human teachers, Tan compares the capabilities of a digital Confucius with those of a human teacher across three dimensions: "teaching the classics and six arts; catering one's teaching to suit the learner's profiles and contextual needs; and being an ethical-spiritual guide and role-model" (Tan, 2020, pp. 282–283). In each of these dimensions, Tan argues that while digital Confucius could serve as a valuable partner to human teachers in transmitting knowledge and skills, it would be deficient in several areas, particularly in functioning as an ethical-spiritual guide and role model for students. In essence, A.I. teachers cannot fully replace human teachers because A.I. teachers do not have what Confucian calls "heart-mind (*Xin* 心)," which is a "distinctly Confucian concept where the cognitive and affective domains are integrated" (Tan, 2020, p. 288). Therefore, A.I. can be a partner to human teachers, which "could be appropriately utilised to complement the work of the Confucian teacher in advancing ethical and spiritual tenets and practices" (Tan, 2020, p. 289), but A.I. can never become a Confucian teacher.

This topic presents an intriguing avenue for future research. For the purposes of this chapter, it is essential to highlight the significant point emerging from this discussion. The point of note is the alignment between Tan's discussion on Confucianism and A.I. teachers, as well as the earlier exploration of the Confucian intimacy account and technology ethics in education. Similar to the considerations surrounding Zoom or i-Beacon technology, the introduction of a digital Confucius as a teacher also raises concerns about privacy and intimacy. When students have a greater familiarity with their human teachers, they tend to feel more confident in using the technology without apprehensions about privacy issues. However, if the teacher is an A.I., limitations on trust may arise, making it more challenging to establish a close teacher-student relationship. This aligns with Tan's ideas that while a digital Confucius can serve as a helpful A.I. partner, it may struggle to handle the nuances of the teacher-student relationship, especially in areas where it is expected to function as an ethical-spiritual guide and role model for students. Digital Confucius faces challenges in fulfilling these roles without the presence of the heart-mind (*Xin*心). These challenges arise from the inherent limitations of A.I. teachers in establishing meaningful connections and intimate relationships with students, as emphasized by the Confucian intimacy account. This demonstrates the importance of recognizing the nuances associated with privacy and intimacy when integrating A.I. teachers into educational settings. Further investigation is warranted to explore how Confucian principles and educational ethics can inform the development and implementation of A.I. teachers, addressing the privacy and intimacy concerns that arise in the context of their usage.

Both Elder and Tan, although not explicitly focused on privacy concerns within their respective topics, have delved into the intricacies of Confucianism's emphasis on relationships and its practical applications in A.I. and technology ethics. By combining these ideas with the privacy issues discussed earlier, particularly the Confucian intimacy account, we can gain a deeper understanding of how Confucianism's distinct perspective on relationships can assist in the debates on the ethics of privacy. Moreover, it is important to avoid viewing this as a simplistic East-West dichotomy. Instead, we should recognize that Confucianism's contributions to these debates are unique and applicable regardless of geographical boundaries.

In conclusion, the discussion in this chapter provides a preliminary understanding of Confucianism's potential contribution to addressing privacy concerns. Further research is warranted to explore these topics further and fully uncover the extent of Confucianism's relevance in addressing real-world privacy issues in A.I. and technology ethics.

Acknowledgements This chapter is supported by the Teaching Development Grant at Hong Kong Baptist University, titled "Ethics of Privacy in Using Technologies for Engaging Teaching and Learning Experiences in Virtual and Physical Classrooms" (No. TDG/2223/04).

References

Allen, A. L. (2011). *Unpopular privacy: What must we hide?, Studies in feminist philosophy*. Oxford University Press.

Carleo, I. I. I., & Robert, A. (2021). Confucian freedom: Assessing the debate. *Asian Philosophy, 31*(3), 211–228. https://doi.org/10.1080/09552367.2021.1899439

Chan, B. S. B. 陳成斌. (2020). A preliminary philosophical investigation into moral issues of the COVID-19 pandemic 2019冠狀病毒病疫情下的道德議題的初步哲學探索. *International Journal of Chinese & Comparative Philosophy of Medicine* 中外醫學哲學, *18*(1), 89–112. https://doi.org/10.24112/ijccpm.181690

Chan, B. S. B. 陳成斌. (2022). A preliminary investigation of privacy issues during the COVID-19 pandemic 初探2019冠狀病毒病疫情下的私隱議題. *International Journal of Chinese & Comparative Philosophy of Medicine* 中外醫學哲學, *20*(1), 63–81. https://doi.org/10.24112/ijccpm.202224

Chan, K. L., Chin, D. C. W., Wong, M. S., Kam, R., Chan, B. S. B., Liu, C.-H., Wong, F. K. K., Suen, L. K. P., Yang, L., Lam, S. C., Lai, W. W.-l., & Zhu, X. (2021). Academic discipline as a moderating variable between seating location and academic performance: Implications for teaching. *Higher Education Research & Development*, 1–15. https://doi.org/10.1080/07294360.2021.1928000

Chan, K. L., Song, X., Kwok, C. Y. T., Kam, R., Chan, B. S. B., Liu, C.-H., Wong, F. K. K., & Wong, M. S. (2022). COVID-19 and contactless learning and teaching: The impact of active participation and user acceptance. In C. Hong & W. W. K. Ma (Eds.), *Applied degree education and the future of learning* (pp. 439–460). Springer.

Elder, A. (2020). Conversation from beyond the grave? A neo-confucian ethics of chatbots of the dead. *Journal of Applied Philosophy, 37*(1), 73–88. https://doi.org/10.1111/japp.12369

Francis, L., & Francis, J. (2017). *Privacy: What everyone needs to know*. Oxford University Press.

Gerstein, R. S. (1978). Intimacy and privacy. *Ethics, 89*(1), 76–81. https://doi.org/10.1086/292105

Gilligan, C. (1982). *In a different voice: Psychological theory and women's development*. Harvard University Press.

Griffiths, S., Man Wong, S., Coco Kwok, Y., Roy Kam, C., Lam, S., Lin Yang, L., Yip, T., Heo, J., Chan, B. S. B., Xiong, G., & Keru, L. (2019). Exploring bluetooth beacon use cases in teaching and learning: Increasing the sustainability of physical learning spaces. *Sustainability, 11*(15). https://doi.org/10.3390/su11154005

Hongladarom, S. (2016). *A Buddhist theory of privacy*. Springer.

Inness, J. C. (1992). *Privacy, intimacy, and isolation*. Oxford University Press.

Kahane, G., Everett, J. A. C., Earp, B. D., Caviola, L., Faber, N. S., Crockett, M. J., & Savulescu, J. (2018). Beyond sacrificial harm: A two-dimensional model of utilitarian psychology. *Psychological Review, 125*(2), 131–164. https://doi.org/10.1037/rev0000093

Lam, C. S. Y., Wong, M. S., & Chan, B. S. B. (2022). Big data, algorithms and whole-person education—the ethical challenges of privacy issues on using iBeacon technology in higher education in Hong Kong. In B. S. B. Chan & V. C. M. Chan (Eds.), *Whole person education in East Asian Universities: Perspectives from philosophy and beyond* (pp. 225–239). Routledge.

Li, C. (1994). The Confucian concept of Jen and the feminist ethics of care: A comparative study. *Hypatia, 9*(1), 70–89.

Li, C. (2002). Revisiting Confucian Jen ethics and feminist care ethics: A reply to Daniel Star and LijunYuan. *Hypatia, 17*(1), 130–140.

Li, C. (2014). The Confucian conception of freedom. *Philosophy East & West, 64*(4), 902–919. https://doi.org/10.1353/pew.2014.0066

Li, C. (2022). The confucian ren and care debate: Reassessment, development, and future directions. *Philosophy Compass, 17*(9), e12868. https://doi.org/10.1111/phc3.12868

Lü, Y.-H. (2005). Privacy and data privacy issues in contemporary China. *Ethics and Information Technology, 7*(1), 7–15. https://doi.org/10.1007/s10676-005-0456-y

Lundgren, B. (2020). A dilemma for privacy as control. *The Journal of Ethics, 24*(2), 165–175. https://doi.org/10.1007/s10892-019-09316-z

Ma, Y. (2019). Relational privacy: Where the East and the West could meet. *Proceedings of the Association for Information Science and Technology, 56*(1), 196–205. https://doi.org/10.1002/pra2.65

Macnish, K. (2018). Government surveillance and why defining privacy matters in a Post-Snowden world. *Journal of Applied Philosophy, 35*(2), 417–432.

Mainz, J. T., & Uhrenfeldt, R. (2022). Privacy rights, and why negative control is not a dead end: A reply to Munch and Lundgren. *Res Publica, 28*(2), 391–400. https://doi.org/10.1007/s11158-021-09524-1

Menges, L. (2021). A defense of privacy as control. *The Journal of Ethics, 25*(3), 385–402. https://doi.org/10.1007/s10892-020-09351-1

Mihailov, E. (2023). Measuring impartial beneficence: A kantian perspective on the Oxford utilitarianism scale. *Review of Philosophy and Psychology, 14*(3), 989–1004. https://doi.org/10.1007/s13164-021-00600-2

Mill, J. S. (1859). On Liberty. https://en.wikisource.org/wiki/On_Liberty
Mills, K. (2022). Consent and the eight to privacy. *Journal of Applied Philosophy, 39*(4), 721–735. https://doi.org/10.1111/japp.12592
Munch, L. A., & Mainz, J. (2023). To believe, or not to believe—That is not the (Only) question: The hybrid view of privacy. *The Journal of Ethics*. https://doi.org/10.1007/s10892-023-09419-8
Noddings, N. (2013). *Caring: A relational approach to ethics and moral education* (2nd ed.). University of California Press.
Nussbaum, M. (2003). What's privacy got to do with it? A comparative approach to the feminist critique. In S. A. Schwarzenbach & P. Smith (Eds.), *Women and the United States constitution: History, interpretation, and practice* (pp. 153–175). Columbia University Press.
Rachels, J. (1975). Why privacy is important. *Philosophy & Public Affairs, 4*(4), 323–333.
Reiman, J. H. (1976). Privacy, intimacy, and personhood. *Philosophy & Public Affairs, 6*(1), 26–44.
Solove, D. J. (2008). *Understanding privacy*. Harvard University Press.
Star, D. (2002). Do confucians really care? A defense of the distinctiveness of care ethics: A reply to Chenyang Li. *Hypatia, 17*(1), 77–106.
Tan, C. (2020). Digital Confucius? Exploring the implications of artificial intelligence in spiritual education. *Connection Science, 32*(3), 280–291. https://doi.org/10.1080/09540091.2019.1709045
Thomson, J. J. (1975). The right to privacy. *Philosophy & Public Affairs, 4*(4), 295–314.
Wang, Q. 王强芬. (2019). The modern value of Confucian ethics for construction of privacy ethics in the big data era [儒家伦理对大数据隐私伦理构建的现代价值]. *Medicine and Philosophy 医学与哲学, 40*(1), 30–34.
Whitman, C. B. (1985). Privacy in Confucian and taoist thought. In D. Munro (Ed.), *Individualism and holism: Studies in Confucian and taoist values* (pp. 85–100). University of Michigan, Center for Chinese Studies.
Yuan, L. (2002). Ethics of care and concept of Jen: A reply to Chenyang Li. *Hypatia, 17*(1), 107–129.
Zimmer, M. (2010). "But the data is already public": On the ethics of research in facebook. *Ethics and Information Technology, 12*(4), 313–325. https://doi.org/10.1007/s10676-010-9227-5

Chapter 3
Xerox and Jixin: On Metonymy and Large Language Models

Levi Checketts

Abstract Current Large Language Models have stirred controversy about the potential of machines to replicate human thought. I argue that such claims are rooted in a highly limited understanding of human cognition, one rooted in a reduction of human language to the electronic replication of English letters. Following the work of Jacques Lacan, I contend that we read LLM generations as meaningful because we human beings are "thrown" into meaning-making. Through our use of symbolic language, we interpret the generation of signs from a machine to be meaningful communication. But metonymic thinking helps us escape this trap by showing that meaning itself is not symbolic but rather imaginary. The untranslatability of Daoist concepts like *Dao*, *jixin* and *wuwei* clarify deeper meaning beyond relations of symbols.

Keywords Large Language Models · Meaning-making · Jacques Lacan · Metonymy · Daoism

3.1 Excitement Around Large Language Models

The beginning of the 2020s saw public interest in Large Language Models (LLMs) reach fever pitch. Mo Gawdat claimed that Google was "making God" (Robitzski, 2021), while Blake LeMoine (2022) said Google's LLM was already "sentient." Yuval Harari, Tristan Harris and Aza Raskin published a letter in the *New York Times* about why AI might wipe us out (2023), while around the same time, Richard Nixon's Secretary of State Henry Kissinger wrote a book with former Google CEO Eric Schmidt and MIT professor Daniel Huttenlocher on the same topic (2021). On

L. Checketts (✉)
Department of Religion and Philosophy, Centre for Applied Ethics,
Hong Kong Baptist University, Kowloon, Hong Kong
e-mail: checketts@hkbu.edu.hk

L. Checketts, B. S. B. Chan (eds.), *Social and Ethical Considerations of AI in East Asia and Beyond*, Philosophy of Engineering and Technology 47,
https://doi.org/10.1007/978-3-031-77857-5_3

the other hand, the news is filled with any number of stories publicly proclaim AI to be an important (possibly *the* important) tool of the future. Public statements emphasize the need for good AI ethics, but then large tech companies like Google (Schiffer, 2021) and Microsoft (Schiffer & Newton, 2023) fire their ethics teams. Some people are emphasizing the need to recognize right for AI programs, while others warn that confusing humans with machines could lead to disaster. And over all of this, people who have made billions off of AI, such as Elon Musk (Future of Life Institute, 2023) or Sam Altman (Center for AI Safety, 2023), have published open letters urging *other* companies not to make LLMs like the ones they have.

It's very clear at the very least that there *is* a lot of excitement. Elsewhere, I have contended in a deeply cynical way that the excitement is meant primarily to generate public interest and thereby more profit (Checketts, 2024). For the purposes of this chapter, however, I will treat the excitement on its own terms. But what is the excitement about? Generally, the fuss is about what AI is about, that is, "thinking machines." The impressive outputs generated by new chatbots are leading people from all walks of life to discuss whether machines are intelligent and whether that has a moral meaning to it. This is, of course, an old discussion, dating at least as far back as Hubert Dreyfus's seminal work on artificial intelligence in 1972 (1992). I am not adding much to the conversation today, except to note how limited a perspective it is to proclaim an advanced calculating machine to be "intelligent" the way human beings are. "Intelligence" is far too fraught of a word to ably discuss in this brief chapter. Rather, I inquire into the question of meaning-making. Since LLMs are supposed to generate meaningful communication, it is important to ask if a machine *can* make meaning at all.

The framework in which I analyze the sense of meaning making in LLMs is psychoanalysis, specifically Jacques Lacan's discussion of modes of thought. For Lacan, our three modes of encountering the world, three "registers of thought," are the real, the symbolic and the imaginary (Lacan, 2006b). Within child development, we migrate from the real to the imaginary to the symbolic. This migration is tied to development of language and concept. The pre-linguistic infant encounters "the real," a world where there is no symbol, there only is what exists. As language develops, the child remains in this mode of thought until they begin to develop an understanding of the gap between word and world. This begins with the process of privation, such as the mother being absent. Privation leads to frustration when the infant finds it cannot find satisfaction as there is no perfect union. For a few formative years, the child exists in the imaginary stage then, where word does not match reality, but also does not have a clear corollary (Lacan, 2020). It is in the imaginary stage, not the symbolic, where language gains "its weight, its resources and its emotional vibration" (Lacan, 1991, 306). In the imaginative stage, language has a "metonymic" form and words exist beyond themselves, whereas in the real, language functions as identity and words function as signs but not symbols. The imaginary stage is resolved through "castration," a conclusion to the Oedipus complex wherein the child submits themselves to the rule of law (or the name of the father) (Lacan, 2020). The child then uses language in a symbolic way; language occupies a metaphoric function wherein words and concepts relate to each other under a system of rules. Following the thought of Jacques Lacan, I argue that LLMs cannot emerge into the symbolic and thus cannot make meaning.

The structure of AI and the assumptions underlying its epistemological model demonstrate a highly limited perspective about thought and language. Monolingual English speakers, coming out of traditions of cryptography and cybernetics, took their experience of language as sign manipulation to be indicative of the entirety of human thought. Following the work of John Searle, I call this the Xerox argument and contend that this form of language is "identity" language, where word and reality correspond exactly. From a strictly identity-thought model, LLMs cannot be said to make meaning as they are different from us, but within a metaphorical-symbolic structure—the ordinary function of human language—we can discuss whether AIs are conscious "like us." Lacan argues that meaning comes from beyond this, however—that something pre-symbolic gives "weight" to the symbols themselves. As a way beyond this form of thinking, I turn to insights from Daoism, especially the discussion of *jixin* in the *Zhuangzhi*. *Jixin* demonstrates for us the problem of identity thought, especially if language is reduced to its alphabetical structure as computer programming must do. Furthermore, Daoism emphasizes the non-exhaustive or non-reducible nature of things, posing the challenge of non-symbolic thought. I thus conclude that metonymic expressions, the sort of thinking that falls within the "imaginary register" of Lacan, transcends the binary conscious/non-conscious question of AI while simultaneously emphasizing the psychotic nature of AI thought.

3.2 Identity Thought and Xerox

LLMs have developed into impressive performance with what they can generate. The way these programs generate plausible-sounding responses to given inputs, responses which we can read and interpret in a meaningful way, has led to discussions about how "smart" the machines actually are. The most extreme interpretations are that the LLMs are "sentient" or "conscious" or some other variation of higher cognitive functioning which we ordinarily ascribe to moral agents such as ourselves (Gunkel, 2023). As such, some people have invoked John Searle's "Chinese Room" thought experiment from 1980 to discuss how we should think of the machine. The thought experiment is as follows:

> Suppose that I'm locked in a room and given a large batch of Chinese writing. Suppose furthermore… that I know no Chinese, either written or spoken…. To me, Chinese writing is just so many meaningless squiggles. Now suppose further that after this first batch of Chinese writing I am given a second batch of Chinese script together with a set of rules for correlating the second batch with the first batch. The rules are in English, and I understand these rules as well as any other native speaker of English. They enable me to correlate one set of formal symbols with another set of formal symbols, and all that 'formal' means here is that I can identify the symbols entirely by their shapes. Now suppose also that I am given a third batch of Chinese symbols together with some instructions, again in English, that enable me to correlate elements of this third batch with the first two batches, and these rules

> instruct me how to give back certain Chinese symbols with certain sorts of shapes in response to certain sorts of shapes given me in the third batch. Unknown to me, the people who are giving me all of these symbols call the first batch "a script," they call the second batch a "story," and they call the third batch "questions." Furthermore, they call the symbols I give them back in response to the third batch "answers to the questions." and the set of rules in English that they gave me, they call "the program." (Searle, 2004, 236–237).

Searle goes on to emphasize that nobody would realistically call this transcription of symbols "consciousness." It is clear that the hypothetical transcriber in this scenario—a copyist less adept than semi-literate medieval monks—has no understanding of what they are doing. Searle's thought experiment has been rather convincing because he gives us what I call the "Xerox" argument. Just as a Xerox machine copying someone's writing is not itself generating meaningful communication, the reproduction of answers itself cannot be "thinking" (cf Lacan, 1991, 304). But this is what the "LLMs may be conscious" folks are saying: that if the outputs of the machine resemble those that might be expressed by thinking beings, then there is no reason to assume the machine is not thinking. Indeed, this was Alan Turing's main argument in "Computing Machinery and Intelligence" (1950): that if a computer could produce answers to queries such that an arbiter would reasonably assume the answers came from a thinking person, then the machine can be said to be thinking. Many AI researchers, from Alan Turing to John McCarthy, have said that they are not interested in the question of consciousness—they consider this to be irrelevant to the question of intelligence or thinking. One should note, then, that this highlights the ambiguity regnant in the discussions of intelligent machines. Despite the fact that AI researchers are emphatic about the need for precise models for their machines, some do maintain that consciousness is merely a pattern in the brain, and others say machines are becoming "sentient." The imprecision employed in discussing what cognition is hints toward the reality that there are various ways we think, and various ways to discuss what thinking is. Indeed, quite recently, a long-standing bet between philosopher David Chalmers and neuroscientist Christof Koch was settled in favor of Chalmers when, after 25 years, scientists were still unable to provide a clear neuroscientific account of what or where consciousness is (Horgan, 2023).

To see why "meaning-making" is a more important criterion, consider Searle's own shortcoming. While he demonstrates well the problem of assuming that AI is conscious, his thought experiment demonstrates a confusion about what happens in thinking. Thinking is not a question of accurately producing the right answer to given questions. A person can be conscious or understanding even without yielding an answer we consider "correct." I can deliberately misanswer a question, can feign ignorance, can provide an indirect answer, can change the subject or merely ignore the question itself. To frame understanding in such polarized terms—yielding *the* correct answer or not—already reduces meaning to cybernetic feedback. This is without even considering that consciousness or understanding varies. Searle's explanation that he is entirely ignorant of Chinese demonstrates the problem that many people who are not monolingual will have varying degrees of understanding other languages. That they may be able to answer some questions correctly, may misunderstand others, and will be entirely ignorant on the meaning of others still. This problem of reducing understanding into monolingual question-answer experience demonstrates a major problem of AI theory of mind.

Searle maintains the underpinnings of the Xerox argument regarding meaning-making, a view that treats our use of language primarily as identity. In the language as identity framework, words and concepts correspond to a single, or a limited, set of specific realities. Language cannot exist in metaphor, where ambiguity and representation lie; it exists rather in the realm of the "real." In this view, A = A, that is to say, there exists no distinction between what is said and what is real. Language is thus understood as universal, and what is said in one context or setting can be understood in entirely the same way as intended in another setting, even when the language must be translated. Indeed, Searle's assumptions about what Chinese language is or is not treats thought as though it were a matter of cryptography—translation is reduced to transposing. A more careful examination of this assumption, which is the same as that of the AI pioneers, reveals that this epistemology reduces human cognition to mathematics. "In cybernetics, the notion of the message has nothing in common with what we usually call a message, which always has a meaning. The cybernetic message is a sequence of signs. And a sequence of signs always comes down to a series of 0s and 1s" (Lacan, 1991, 304).

Turing's ideas about the "thinking" potential of computers derived from his initial work of creating sophisticated adding machines for cryptography, taking an alphabetical communication and converting the letters into other symbols. This approach reduces language to its smallest written component and ignores broader questions of semantics, grammar, pronunciation, and so forth. What emerges is a language of syntax without semantics (Lacan, 1991): formal rules without meaning. Thus, language is entirely reduced to signs: in the case of Turing and his World War II work, the Latin alphabet. One can represent the English/German alphabet which Turing was decoding with 26 letters. Thus, one only needs a total of 26 individual symbols (more, if you include punctuation) to represent all literal English or German thought. Nor are English and German unique as many other languages also use a relatively small set of letters, such as Greek with 24, Arabic with 28, Hebrew with 22. One can learn to read and write any of these within a few hours. Once one has learned the alphabet, they can then write or read anything, though as Searle clearly notes, this does not mean one understands.

Turing's work involved a mechanical calculator which had switches that could be set in discrete positions of on or off. This is the basis for all *digital* technology—discrete states situate both the inputs and the outputs. Thus, the history of AI has to be traced to telegraphy, a system where all written communication is conveyed through "the fairy electricity…[establishing] circuits which open and close, which interrupt themselves or restore themselves, as a function of the existence of cybernetized doors" (Lacan, 1991, 302). Since on and off constitute two discrete states, switches (or electrodes today) constitute binary machinery, with a 0 meaning off and a 1 meaning on. With a row of switches, one can then program simple binary patterns. The twenty-six letters of the English alphabet can be encoded in binary with just five switches or digits—11010. In standard computing, each letter will require a total of five bits—binary digits—to encode. If one includes all ninety-five inputs for a standard English keyboard, including numbers, punctuation and capitalization, the number goes up only to seven binary digits—1011111, bringing the number up to seven bits. A byte is the most common measurement of computer memory and consists of eight bits, enough to denote any input from a standard keyboard. For early cybernetics, however, numbers and punctuation were not in use, leaving essentially the limit of twenty-six letters, signs that had long-since been conveyed electronically though Morse code.

Turing's approach to communication reduces language not only to the word, but more significantly to the letter, the pure signifier (Lacan, 2006a, b, 23, 42–45). Chinese, on the other hand, is not so reducible. Chinese is not written in letters. The *hanzi* are syllables, and there are over 50,000 Chinese *hanzi*, though most people only know about 8000. All of these 50,000 *hanzi* would require a total of sixteen binary digits—1100001101010000. This is several orders of magnitude greater than the Latin alphabet. Moreover, Chinese has a history of other obstacles in electronic transcription on a computer screen. ASCII, the standard form for encoding characters in computers, uses just 475 bytes to encode the entire English writing system. Chinese, which has more complex characters than English letters and far more of them, should require at least 256 kilobytes, or 551 times more memory. Indeed, a reliable Chinese word processing program did not emerge for computers until the mid-80s after some very clever computer programming found ways to get around the problem of too much memory requirement (Mullaney, 2021).

There are ways of ordering and organizing *hanzi*, of course. The *Kangxi Dictionary* uses 214 *Kangxi* radicals as root characters for organizing all Chinese words. A 214 "radical" ordering system is itself much more unwieldy than a 26-letter alphabet, and it becomes even more complicated given that many *hanzi* are composed of multiple radicals. But it is here where the relationship becomes even more complicated—the meaning of a given *hanzi* can be derived from a symbolic connection between radicals, or, more often, through a metonymic one (Lacan, 2011, 88). Thus, Chinese etymology in the *Shuowen Jiezi* articulates six ways radicals are connected to form new words. This complex system clearly can be learned and serves a massive portion of the human population well, and undoubtedly operates within the symbolic order. However, it does not lend itself well to the binary reduction of language to pure symbols that can be abstracted the way Turing et al. did. Indeed, the most pressing need to do this, within the science of telegraphy as it was for English language, resulted in a four-digit notation system, with the first two digits noting pages of a 100-page volume, and the second two locating the character on a 10 by 10 graph on each page. There is also the "four corner" system used to note different characters for each of the "four corners" of a Chinese *hanzi*. In contrast, Morse Code classifies all 26 letters with four sequences only, and can be learned by even cub scouts relatively quickly.

On all fronts, then, we can see a sharp qualitative difference between the alphabetical writing style of English and the character system of Chinese. As English written communication was formalized and standardized through print, typewriting, and telegraphy, it could be easily transmitted through electric signals and reduced to binary inputs with little complication. Insofar as the English letter could be reduced to a mere sign and abstracted from any and all meaning, it could be reduced to a different sign or even a sequence of strokes (Lacan, 2011). Of course, the *meaning* of these strokes still requires human understanding (Lacan, 1991, 2006a). And while one can theoretically do this with Chinese, just as one can strip the Latin alphabet of its genealogical symbolism, it remains technically much more difficult as one can see even in the difficulty of creating a Chinese typewriter (Mullaney, 2018). In short, the simplistic assumption that language can be reduced to something *programmable* into a binary machine may appear readily to persons who conflate language and communication to code, but from a Chinese perspective, written language is not so reductive.

3.3 From Identity to Metaphor

LLMs' mode of "thought" is stuck in language as identity. The elements of "thought" are binary sequences, ones and zeroes programmed by mathematicians who, following the view of Leibniz, assume that all human thought is restricted to a limited (if large) set of symbols. Words have set meanings and follow strict rules in grammar, and these words are themselves made up of discrete elements called letters. Any deviation from the established norm is a failure to communicate. Thus, even the highly technical non-flexible instructions used to program a computer are called a "language," and any incorrect instruction—such as a misspelling, missed punctuation, or a missing command—will result in the program not functioning properly. This reduction of language and thought infects both the overly mathematized view of the AI apologists and people like Searle who critique the specific contours, not the general thesis of AI epistemology. According to this view, whether articulated through Turing or Searle or contemporary AI evangelists, language admits of one sense. There is no need for hermeneutics, and there is no room for ambiguity. Even the contestable question of whether LLMs are conscious *assumes* that consciousness means *one* thing, a binary wherein an entity is *either* conscious (1) or not (0).

Ironically, it is only because we typically think in *metaphoric* modes rather than *identical* modes that we can accept the proposition that a machine is intelligent *like us*. We call the non-communicational mode of computer programming a "language" because we see a resemblance to real human language. This resemblance, Lacan argues, is due to the fact that machines do, in fact, employ syntax. Speaking positively of the nascent work on AI, Lacan noted "cybernetics is a science of syntax, and it is in a good position to help us perceive that the exact sciences do nothing other than tie the real to a syntax" (Lacan, 1991, 305). However, he goes on to note "it is us who introduce meaning" which a machine cannot do because "nothing unexpected comes out of the machine" (ibid.). We interpret meaning-making as coming from behind the discourse of logic through "the symbolic function which is manifested through it" (306).

The symbolic is the realm in which we encounter signifiers, while signs exist for non-symbolic creatures, such as dogs or machines. The difference between sign and signifier demonstrates the complicated nature of human thought. The sign, Lacan says, "is to represent something for someone" (Lacan, 2011, 36). The signifier, on the other hand, "represents the subject for another signifier" (Lacan, 2011, 37). This confusing distinction is made apparent in the phenomenon of identification itself—the way in which we identify one signifier with another, such as my therapist with my father in transference, a rock formation with a lion in pareidolia, or Searle's "squiggles" with a written language in reading. Lacan notes that only communicating beings, that is subjects, use signifiers. Other beings, including dogs which have their own markers and pre-verbal speech, use signs. At best, computers operate with signs: they are incapable of the sort of interpretative flexibility that is required for the use of signifiers and thus metaphorical language. "With a machine, whatever doesn't come on time simply falls by the wayside and makes no claim on anything. That is not true for [the human], the scansion is alive, and whatever doesn't come on time remains in suspense. That is what is involved in repression" (Lacan, 1991, 308). In the "Seminar on the 'Purloined Letter'," Lacan notes that the letter is a *pure* signifier and that it "always arrives at its destination" (Lacan, 2006a, 30). We human beings, who are "thrown into [meaning], committed, caught up in its gears" (Lacan, 1991, 307), cannot help *but* see meaning in the signs while non-repressed entities (like dogs and computers) merely formulate signs.

In approaching language as metaphor, we recognize the typical relationship between sign and signifier—the word (written or spoken) points to something beyond it. The word is not the thing itself. AI epistemology cannot grapple with this—the word *is* the thing. An algorithm relies on strict rules, where outcomes are givens for inputs. The English word, reduced to first letters, then codified into ones and zeroes, *is* what the LLM computes; it does not comprehend, reflect, suppose, or understand. And certainly it does not make meaning. It merely computes a mathematical response. But when we use language, we usually use it in a metaphorical way. Consider the use of "we" in the previous sentence: I (an early twenty-first-century scholar designated Levi Checketts) invoke this term to include my hypothetical readers (you, both my implicit and real readers). Together, you and I constitute "we." I do not know who will read this, and so cannot make a clear statement about who "we" constitutes (and you may not know "me" either beyond the trait that designates me through a proper name). But human beings (again, drawing a generalization through metaphor) use symbolic ways of thinking to understand words in a more general sense. Thus, Lacan argues that we should never think that A = A (Lacan, 2011). The reason we should not say this is because in delineating the first A and the second, we are implying some difference. "I am a man" is not a tautology but rather a practice of identification, perhaps because someone has challenged my masculinity or I am introducing myself or commiserating with other men. The computer, however, accepts A = A as tautological because it operates merely on identity thought. A is A without loss or transfer in computer thinking; it need not be established that A is A in fact nor is it in question because the computer does not exist within the symbolic order. The computer is in the real—things are always where we find them, and so A is always A.

Thus, when the conversation surrounding AI turns to its consciousness or sentience, we are using metaphorical language. We assume the computer can or cannot be "like me"—we assert the signifier "consciousness" for other signifiers to establish that what the computer does is what we do. Then, the conversation continues through metaphor as we use signifiers to designate a "signified" behind them. Computers, however, merely *produce* signs, which we might interpret as signifiers. They do not create meaning, but we, who have meaning given to us, are likely to read their outputs as meaningful. The purity of mathematical reduction allows the machine to, at best, function with identity logic, which is *not* inherently meaningful because it lacks both an intentional connection to other symbols and the weight given by the imaginary.

The metaphoric way of thinking is so common that we often forget that it *is* metaphoric. A word like "man" easily designates a male adult, general humanity (though this is archaic), or even a non-descript authority (*the* man), and can easily be adapted in ways that demonstrate interpretative flexibility, such as various superheroes having "man" in their name including the alien Superman, or Tolkien's famous subversive prophecy in *The Lord of the Rings* (that "no man" could kill the Witch King, but a *woman* could). Beyond this, of course, visual symbols like flags or national emblems, the function of repression, and modes of operation in various settings demonstrate the importance of symbolic representation. Indeed, one of the persistent problems vexing AI today is its inability to adapt to novel situations, a task which we typically do well because of our own ability to interpret flexibly.

Going beyond the mathematical transcription of Latin-based written languages also helps to demonstrate metaphorical thinking. Translation, not transcription, requires metaphorical thought. Words do not have the exact same meaning across languages, and so translators are forced to choose words and expressions *not identical*, but most *appropriate*. Jacques Derrida notes that the task of translation is one that is inherently violent, as the translation is never exactly the same as what the original author meant. Thus, a translator carries a great responsibility and must take ownership of the translation (Derrida, 1985). This fittingness of a translation is a question of metaphor—does the word or phrase chosen to render another idea *fit* the sense of the initial statement? What does "fitting" mean? The field of hermeneutics has long demonstrated the complexity of this. History, cultural context, author, audience, and linguistic nuances are all critical for translating thought into something we might consider at least passably accurate.

Drawing further insights from Chinese, consider this word: 機心 (*jixin/geisam*). The first character 機 combines two characters: 木, mu/mok (meaning wood), and 幾 ji/gei (meaning how many). These are both common characters, but even the second one is comprised of two other characters: 幺 yu (small), and 戍 shu/syu (defend). None of this is important to understand that 機 itself means "spinning machine" or "weaving machine" initially and "machine" more generally. How does one get "machine" out of "wood/how many" or "wood/small/defense"? Knowing how to *write* and *read* the character requires understanding that there are these various components which together make up this *one* character, which means something different from, but etymologically inheritor of the other *hanzi*. And, indeed, if one simply misses the 木, then the meaning of the word is entirely different. The second character, 心 , is much simpler and means heart/mind. The fact that it means both is important for us to see why rote transcription does not amount to translation, why identity is a bad model for thought. Translated as directly as we can (if that even makes sense), the full word *jixin* means "machine mind/heart," a fitting notion for AI, I think, but that is not what the word typically means.

To see the meaning of the term, consider this passage from *Zhuangzi*.

> Zi-gong had been rambling in the south in Chu, and was returning to Jin. As he passed (a place) on the north of the Han, he saw an old man who was going to work on his vegetable garden. He had dug his channels, gone to the well, and was bringing from it in his arms a jar of water to pour into them. Toiling away, he expended a great deal of strength, but the result which he accomplished was very small. Zi-gong said to him, 'There is a contrivance here, by means of which a hundred plots of ground may be irrigated in one day. With the expenditure of a very little strength, the result accomplished is great. Would you, Master, not like (to try it)?' The gardener looked up at him, and said, 'How does it work?' Zi-gong said, 'It is a lever made of wood, heavy behind, and light in front. It raises the water as quickly as you could do with your hand, or as it bubbles over from a boiler. Its name is a shadoof.' The gardener put on an angry look, laughed, and said, 'I have heard from my teacher that, where there are ingenious contrivances (機械), there are sure to be subtle doings (機事); and that, where there are subtle doings, there is sure to be a scheming mind (機心). But, when there is a scheming mind (機心) in the breast, its pure simplicity is impaired. When this pure simplicity is impaired, the spirit becomes unsettled, and the unsettled spirit is not the proper residence of the Dao (道). It is not that I do not know (the contrivance which you mention), but I should be ashamed to use it.' (Zhuangzi, n.d., "Heaven and Earth" 11)

The translator has rendered 機心 (*jixin*) as "scheming mind." In connection with "ingenious contrivances" and "subtle doings," one has the sense that *ji* (機) refers to something like the English term "crafty." *Jixin*, the condition where this craftiness conditions the heart/mind, is opposed to the simple way that Daoism advocates for its adherents. The machine and the machine heart are not the ways of Dao. One must understand the relation between a machine (機), the scheming mind (機心) and the Dao (道). Here we are confronted with more non-translatability; even in the "translated" section, we do not translate "Dao." Why not? In most Chinese contexts, the word 道 (*dao/dou*) just means road. My university is located on Baptist University Road, with the Chinese name using 道 for road. In the passage from Zhuangzi, however, the word 道 cannot be treated as a simple substitute for the English word "road." Nobody would say "the unsettled spirit is not the proper residence of the *road*." One *might* read 道 as "way," as in "the unsettled spirit is not the proper residence of the *way*," but this is only possible through a pre-existing metaphorical reading of the word "way." The clunkiness and lost meaning from this literal transcription almost glares at the reader.

Although the passage above demonstrates the function of metaphoric thinking, where we translate *jixin* as "scheming mind," it also demonstrates the limits of symbolic thought. If we take Daoism seriously, the "road" or "way" of 道 ultimately must defy translatability. The *Daodejing*, the founding text of Daoism, opens this way: "The 道 that can be followed is not the eternal 道. The name that can be named is not the eternal name. The nameless is the origin of heaven and earth, While naming is the origin of the myriad things. Therefore, always desireless, you see the mystery. Ever desiring, you see the manifestations. These two are the same—When they appear they are named differently. This sameness is the mystery, Mystery within mystery; The door to all marvels" (Laozi, n.d., 1). In this case, 道 may be rendered as "road" or "way," but if we translate it too literally, we miss the point.

But clearly, this is itself open to skepticism: after all, one only needs to deny the premises of Daoism to assert that the 道 that can be followed **is** the eternal 道. Perhaps a more interesting translation demonstrates the impasse. In Chinese translations of the Bible, this same word 道 is the translation used for the Greek word λόγος (*Logos*) in the prologue of the Gospel of John. In English, we read "In the beginning was the Word (ὁ λόγος, *ho Logo*s) and the Word was with God and the Word was God" (John 1:1) But in Chinese, in the beginning is the 道. Why would the word for "Word" not be the word chosen for Chinese (i.e., 詞)? Because *ho Logos* is not literally "the word" either, as the history of Greek philosophy makes clear to us, and even the meaning of this passage from the Bible suggests this. If one therefore reduces language to letters (and then digital sequences) as Turing and others do, then the literal meaning of John's *Logos* theology is that Jesus is made up of a few letters. Thus, God can truly be a binary sequence as Mo Gawdat and Sam Altman claim. But even a *metaphoric* translation evades us: one must be more attentive to the function of the word within its own linguistic system. Thus, if one understands *Logos* to mean something more like mind, thought or reason, then the translation to "Road" in Chinese makes more sense, given the flexibility of 道 in Daoism. These words carry "weight" in their native contexts which evade straightforward metaphorical translation. Thus, we begin to see as well the importance of considering *metonymic* approaches to language, where a word need not occupy a purely symbolic function but rather a relation to other signifiers, where *meaning* emerges.

3.4 Meaning and the Dao: Metonymy as a Way Out

To some degree, *jixin* represents the worldview of AI. *That* we can create a mind, that our machinations can become "conscious" is an assumption *that* thinking is and should admit of mastery. The push to almightiness, the assumption that we can master "thinking" for example, is an attempt to relieve our fundamental anxiety (Lacan, 2014, 308). However, the secret of anxiety is that it persists. Due to human lack, we can never truly fill the void and never finally master anxiety. The real cannot be contained in the symbolic and so we can never accomplish once and for all the final move that will bring us total satisfaction. There will always be mismatch, misspeaking, and misrepresentation. Thus, "when there is *jixin* (機心) in the breast, [the mind's] pure simplicity is impaired. When this pure simplicity is impaired, the spirit becomes unsettled" (Zhuangzi, n.d., "Heaven and Earth" 11). In contrast to the claims of LLM prognosticators, meaning is not generated by computers because the drive to mastery fails to recognize that meaning itself comes from the weightiness of things beyond our symbolic control of them. Whether one takes this in the form of Lacan's critique of cybernetics or Zhuangzi's critique of ingenious contrivances, the inventor's push for mastery is a vain project that fails to recognize the impasse that defines human existence. Our being is defined as much by non-being and our fullness by lack (Lacan, 2020). Once again, then, Daoism offers a way past this reductive approach to the mind.

The notion of *wuwei* 無為, meaning something like non-action (and like all Daoist notions is strictly impossible to translate into English) is a central moral principle outlined in the Daodejing. The principle emphasizes non-effort, an approach to encountering the world without seeking to control it. As such, the illusion of pure knowledge must be eschewed. "If we could renounce our sageness and discard our wisdom, it would be better for the people a hundredfold. If we could renounce our benevolence and discard our righteousness, the people would again become filial and kindly. If we could renounce our artful contrivances and discard our (scheming for) gain, there would be no thieves nor robbers." (Laozi, n.d., 19). By seeking Dao 道 instead of "objective knowledge" or "wisdom" or "efficiency" or "profit" or any of the dozens of promises made about LLMs, we might avoid mistaking the symbolic-thinking dualism of presenting the identity-thought of LLMs as *either* conscious or not, at least according to a Daoist perspective. Symbolism helps us understand the relationship between benevolence and kindness or gain and thievery, but the irreducibility of 道 gives weight and carries meaning for these conditions.

The non-translatability of 道 offers us an introduction to metonymic thought. In metonymy, the signifier migrates; it is not tied to a set of signifieds. Citing positively the work of post structuralist Claude Levi-Strauss, Lacan uses the structure of myth as a way of demonstrating metonymy (Lacan, 2020). In myth, Levi-Strauss (1955) argues, the relationship of elements to each other is where the meaning lies, not in symbol. Thus, the Oedipal myth is about relationships more than symbolizing anything specific. Mother need not be mother, and it does not have to be a specific symbol; the greater meaning is the relationship mother has. Rather, it is the "weight" these ideas hold, a weight granted by their presence in the imaginary order (Lacan, 1991). The signifiers themselves migrate, but their relationship to each other indicates the greater weight. Thus, the relationship between elements in metonymy and their respective weights takes priority over any symbolism. When Lacan writes about the "phallus," then, he does not mean penis, as many of Freud's interpreters have meant. Rather, the phallus is inherently a lack, something one does not possess or believes that he possesses. Freud's *penisneid* is subverted: the woman's desire for the phallus is a desire for what one lacks, and *if* that lack appears as a penis, it is only to the degree that the phallus, as a migrating signifier, has settled upon the penis (Lacan, 2020). Anxiety as well, which has *nothing* as its object (because it is defined by our lack or incompleteness) often seeks the phallus to bring satisfaction, sometimes in the form of sexual gratification. In the classical family tri-partite structure, though, the child occupies the place of the phallus, at least until the mother recognizes the child does not satisfy desire. Thus, the phallus, as a migrating signifier, at one time means penis, at another means breast, at another still means the infant—it is defined not by its symbolism but by its function in relation to subjects.

The notion of 道 likewise migrates. It can, and frequently does, merely mean road. But the signifier migrates to include a mode of life, or a right mindset. Daoism itself demands that the one to travel this 道 not confine themself to a set of symbols or a structured system. The true 道 must admit of flexibility, migration, openness. It cannot be circumscribed, so it migrates. Thus, Daoism is a prescription for life that requires one to always be attentive, to never assume one has mastery, to pay attention to the finest details while shrugging off weighty matters. From Daoist perspective, the question of consciousness is too weighty to pretend we can settle. The restricted focus of the question "Are LLMs conscious?" restricts consciousness to a set of symbols; it is not the way of Daoism which allows movement and migration. It is rather more important to consider the meaning-making which itself will evade symbolic representation.

But metonymy is not restricted to the Daoist approach. One can look beyond the life following 道 and see the significance of metonymy in disrupting dualistic metaphoric thought or singular identity thought. In narrative, metonymy is demonstrated through the absurd, such as in Kafka's novels, or the films of Terry Gilliam—the signifiers are set but they are not transparent. What they mean is not straightforward. Signifiers move and so meaning can only be understood by tracing the movement of the signifier—not by reducing the elements to pure symbols or explaining the alternative meaning behind them (Lacan, 2011). The absurd is absurd because the elements escape clear signification. The meaning remains on the tip of our tongues, defying expression but clearly there. In such narratives, the machinery of capitalism, the ever-present drone of mass media, the labyrinths of bureaucracies do not *represent* anything (they do not even represent themselves); their function in the setting is what provides their meaning. Likewise surrealist art demonstrates the function of metonymy. Dali's *Great Masturbator* or Magritte's *Personal Values* do not *symbolize* anything. The elements in the paintings relate to each other in ways that defy symbolic communication. The composition of a sleeping face with a locust attached from which a woman emerges near a man's genitals defies ordinary relations between these elements. The "banal" symbolism typically assigned escapes our grasp as a weightier sense of the real seeps in (Lacan, 2011, 41). Zen Buddhism likewise, in its famous *koans*, presents language as non-symbolic but rather metonymic. To ask, "What is the sound of one-hand clapping?" requires the listener to focus on the *relationship* between hands and clapping rather than the *symbol* that clapping may entail. And since koans are often called "riddles" (a poor translation, demonstrating again the failure of metaphor), we might also note that much of *humor* relies on metonymy. To take a simple joke—time flies like an arrow; fruit flies like a banana—defies the analogical (i.e., metaphorical) set up because the signifier "like" migrates from functioning as metaphor to functioning as verb while the signifier "flies" is used in homonymous ways to indicate first movement and second insects. Humor generally functions by interrupting symbolic relations—the buffoonish dignitary, the fiendishly clever rabbit, the socialite's bodily failure, all function as disruptions of the carefully maintained symbolic order.

Each of the cases above, from surrealism to Daoism, illustrate the significance of metonymic expression. Human communication typically lies in the realm of symbolism, including, most obviously, turns of phrase or clichés we employ. It is only because of our capacity for symbolic thought that we can even entertain the notion that LLMs or other AIs might be "conscious." Those programs themselves, though, function through identity thinking—the data input and output are the entirety of the thought. Unlike our symbolic thinking, AI outputs do not point to anything beyond themselves. They do not generate meaning. But we who live in the realm of the symbolic tend to read these as symbols, the same way we might "see" a face on the front of a car or the eponymous beast in Hong Kong's "Lion Rock," (i.e., pareidolia). We who are "thrown into" meaning cannot *but* see meaning where none is intended. The productions of LLMs strike us as communication not because the machine is capable of generating meaning, but because we generate the meaning on behalf of it.

3.5 Conclusion: A Prescription for Imagination

Within the Lacanian framework, our resolution of castration complex determines our typical mode of thinking. Those of us who begrudgingly accept the reality that we cannot have wholeness become neurotics and dwell begrudgingly in the symbolic world. On the extreme end are those who reject the cut, who deny that they are not whole, and who thus become psychotic, taking language as identity. The third option is the pervert; the one who believes they can overcome the gap, whose language is not confined to either identity or metaphor. Law, that is to say, normative systems that govern our encounter of the world, is the major expression of this castration. The neurotic accepts the law, recognizes the need to follow the order and fit into it. Good and bad, correct and incorrect fit within this symbolic system. The psychotic, however, recognizes no law because everything is identity: a law *can* be broken and thus cannot be identical with reality. The pervert, however, recognizes the law but goes beyond it willingly.

We neurotics have a hard time staying within metonymy—we prefer the symbolic order of metaphor, where the sickle means castration and the tower as phallus represents a penis. The system of symbols allows for clear meaning and establishes permissible interpretations over impermissible ones. But even metaphor is beyond the capabilities of language models. Metaphor works because we are capable of *understanding* the signifier is not the signified. There is nothing beyond the sign for the machine—no signified beyond what *we* read into it. The Xerox model of cognition demonstrates this: a given input requires a given output (or at least a limited set of given outputs). Metaphor does require some flexibility; when new images, new concepts or new phenomena enter the symbolic world, they must be accommodated. The symbols can be stretched (and please note that "stretched" here is itself a metaphor—words cannot actually stretch). AI must program in entirely new meanings with set chains of description and explanation for every non-literal use of a term. It may appear to function with metaphor, but really the operation itself is within a set of identities. Any new understanding requires new chains of identity; terms do not exist as symbols.

AI is thus, at best psychotic. It has no symbolic world. It is subject to no law. It cannot because of the way it is designed. We respect it as possibly conscious because we are good neurotics—we read the sign as symbol and thus read in meaning where none was created—the imitations of language through a semantic-less syntax point, we think, to a signified mind. Or, as Lacan himself notes, "the neurotic wants to retransform the signifier into what it is the sign of" (Lacan, 2011, 129). But we need good perverts to challenge this—thought isn't *only* this—it's also nonsense, like supplying fake words, saluting imaginary flags, referring to referents that lack a referent behind them.

References

Center for AI Safety. (2023). *Statement on AI risk.* https://www.safe.ai/statement-on-ai-risk#open-letter

Checketts, L. (2024). *Poor technology: Artificial intelligence and the experience of poverty.* Fortress Press.

Derrida, J. (1985). Des tours de Babel. In J. Graham (Ed. & Trans.). *Difference in translation* (pp. 165–207). Cornell University Press.

Dreyfus, H. (1992). *What computers still can't do: A critique of artificial reason.* MIT Press.

Future of Life Institute. (2023, March 22). *Pause giant AI experiments: An open letter.* https://futureoflife.org/open-letter/pause-giant-ai-experiments/

Gunkel, D. (2023). *Person, thing, robot: A moral and legal ontology for the 21st century and beyond.* MIT Press.

Harari, Y., Harris, T., & Raskin, A. (2023, March 24). You can have the blue pill or the red pill, and we're out of blue pills. *New York Times.*

Horgan, J. (2023, June 26). A 25-year-old bet about consciousness has finally been settled. *Scientific American.*

Kissinger, H., Schmidt, E., & Huttenlocher, D. (2021). *The age of AI: And our human future.* Little, Brown and Company.

Lacan, J. (1991). In J-A. Miller (Ed.), *The seminars of Jacques Lacan II: The ego in Freud's theory and in the technique of psychoanalysis* (S. Tomaselli, Trans.). W. W. Norton.

Lacan, J. (2006a). Seminar on the 'Purloined Letter'. In *Écrits* (B. Fink, Trans.). (pp. 6–48). W. W. Norton.

Lacan, J. (2006b). The situation of psychoanalysis and the training of psychoanalysts in 1956. In *Écrits* (B. Fink, Trans.). (pp. 384–411). W. W. Norton.

Lacan, J. (2011). *The seminars of Jacques Lacan IX: Identification* (C. Gallagher, Trans.). https://www.lacaninireland.com

Lacan, J. (2014). In J-A. Miller (Ed.), *The seminars of Jacques Lacan X: Anxiety* (A. Price, Trans.). Polity Press.

Lacan, J. (2020). In J-A. Miller (Ed.), *The seminars of Jacques Lacan IV: The object relation* (A. Price, Trans.). Polity Press.

Laozi. (n.d.). *Daodejing* (J. Legge, Trans.). Chinese Text Project. https://ctext.org/dao-de-jing

Lemoine, B. (2022). Is LaMDA sentient? — an interview. *Cajun Discordian*. https://cajundiscordian.medium.com/is-lamda-sentient-an-interview-ea64d916d917
Levi-Strauss, C. (1955). The structural study of myth. *The Journal of American Folklore, 68*(270), 428–444.
Mullaney, T. (2018). *The Chinese typewriter: A history*. Oxford University Press.
Mullaney, T. (2021, June 30). The engineering daring that led to the first Chinese computer. *TechCrunch+*.
Robitzski, D. (2021). Former Google Exec warns that AI researchers are 'Creating God'". *Futurism*. https://futurism.com/the-byte/google-exec-ai-god
Schiffer, Z. (2021). Google fires second AI ethics researcher following internal investigation. *The Verge*. https://www.theverge.com/2021/2/19/22292011/google-second-ethical-ai-researcher-fired
Schiffer, Z., & Newton, C. (2023). Microsoft lays off team that taught employees how to make AI tools responsibly. *The Verge*. https://www.theverge.com/2023/3/13/23638823/microsoft-ethics-society-team-responsible-ai-layoffs
Searle, J. (2004). Minds, brains, and programs. In J. Heil (Ed.), *Philosophy of mind: A guide and an anthology* (pp. 235–252). Oxford University Press.
Turing, A. (1950). Computing machinery and intelligence. *Mind, 49*, 433–460.
Zhuangzi. (n.d.). *Zhuangzi* (J. Legge, Trans.). Chinese Text Project. https://ctext.org/zhuangzi

Part II
Christian Dialogues with Culture

Chapter 4
Comparative AI Ethics Between Silicon Valley and the Vatican: Divergent Foundations, Convergent Initiatives, and "How-to" Ideas for Discussing and Developing Technology Ethics

Brian Patrick Green

Abstract The field of AI ethics has gained much attention in recent years, not only as technology has failed in crucial ways, such as spreading misinformation, biasing data evaluations, and even killing people in autonomous vehicle accidents, but also as it has succeeded almost too well, gathering precious data, making huge piles of money, and stealing our attention from our loved ones in order to look at screens. The technology industry has proliferated principles and begun to practice them, while the Roman Catholic Church has wondered what to make of this contemporary technological situation. Several organizations within the Catholic Church have been investigating the ethics of AI from a much deeper philosophical and theological level than that of mere principles and their operationalization (not to malign the clear importance of such practical work). When faced with technologies that can turn society upside-down, corporations don't do philosophy, they do damage control and cost-benefit analysis, while individual people, and particularly religious individuals and religious and educational organizations, should think about these subjects philosophically and theologically. This paper will compare these different approaches, both in terms of what is being done and what should be done.

Keywords Artificial Intelligence · AI Ethics · Technology Ethics · Business Ethics · Religion and Technology · Catholic Church

---

B. P. Green (✉)
Markkula Center for Applied Ethics, Santa Clara University, Santa Clara, CA, USA
e-mail: bpgreen@scu.edu

L. Checketts, B. S. B. Chan (eds.), *Social and Ethical Considerations of AI in East Asia and Beyond*, Philosophy of Engineering and Technology 47,
https://doi.org/10.1007/978-3-031-77857-5_4

4.1 Introduction

In 1959, C.P. Snow described the "Two Cultures" problem between literary intellectuals and physicists (Snow, 1998). Today, the problem has only expanded, and in particular the divide between religion & ethics and corporations & technology is becoming increasingly relevant as artificial intelligence (AI) technology changes the world. AI makes art, converses with us, makes recommendations, offers romantic matches, tracks and evaluates our health, evaluates our worthiness for loans, navigates for us, drives for us, and grabs our attention in any way that it can. AI is ubiquitous, powerful, and in dire need of ethical guidance.

Many technology companies have stepped into this ethical vacuum with principles-based approaches to AI ethics; for example, Microsoft (2023), IBM (2023), Salesforce (2023), and Google (2023). For several years the technology industry was all about formulating principles; now it has shifted to operationalizing principles. The operationalization question is a hard one, since taking the mere words of ethical principles and making them real in human behavior is beyond what most principles-based approaches to ethics actually describe. Operationalization requires thought and dedication, and perhaps even "philosophy" to bridge the theoretical and the practical.

The idea of corporations doing philosophy might remind one of the movie *A Fish Called Wanda*, where Kevin Kline's character Otto (who, due to his ignorance, has been called an "ape" by Jamie Lee Curtis's Wanda) attempts to defend his own sophistication by appealing to his penchant for reading philosophy, declaring to Wanda "Apes don't read philosophy!" But Wanda retorts "Yes they do, Otto, they just don't understand it!" (Cleese & Crichton, 1988).

Corporations as entities do not read philosophy, and as entities they certainly do not understand it. But as entities they do embody philosophies, operationalize philosophies, and commit philosophically-relevant actions. And the humans that compose corporation can, and sometimes do, read (and sometimes understand) philosophy.

However, when faced with technologies that can turn society upside-down, corporate philosophy is likely to be inadequate. This paper will attempt to lay out a few tactics for trying to remedy this deficiency. Individual people in corporations have a role to play, and for the purposes of this paper, religious individuals in corporations and, externally, religious and educational organizations in particular, should think deeply about these technologies philosophically and theologically.

Pope Francis has made known within the Roman Catholic Church that it should be contributing to the social conversation on AI. However, this only raises more questions: What are Catholics, Christians in general, and all people of faith to make of the contemporary technological situation? What contributions might be possible for them to make?

These efforts have already begun. For example, within the Catholic Church several organizations have been investigating AI ethics at a deeper philosophical and theological level than that of mere principles and their operationalization (Green, 2022). These deep theoretical investigations yield clear ethical implications too, and perhaps startingly, many of these are the same ethical ideas that tech companies themselves have reached through their own efforts. What could explain this odd convergence?

This paper will compare corporate and religious approaches to AI ethics, focusing on four technology companies and the Roman Catholic Church. It will explore their philosophical commitments, convergences, and how communication might permit even more to be done. The goal of this paper is not only to inform, but also hopefully to lay out a path forward for greater cooperation on issues related to technology ethics, for the sake of building a better future.

4.2 The Philosophies of Corporations and a Church

4.2.1 Corporations

To begin, it is necessary to know more about the relationship of corporations and philosophy. Certainly, if companies are making ethical AI principles, they must be assuming some foundations for those ethical principles, which indicates some measure of philosophical expression, much of which is implicit.

(1) Adam Smith believed that the division of labor and the concomitant economic activity it generated was for the sake of social benefit, but mediated by self-interest: "It is not from the benevolence of the butcher, the brewer, or the baker, that we expect our dinner, but from their regard to their own interest" (Smith, 1994, p. 15). This philosophy is partially embedded into the contemporary notion of a corporation (though Smith was not well-disposed towards the corporations of his day, p. 790–805). Corporations exist to provide some benefit to society, but do so out of self-interest, and those who engage in economic activity with them agree that the benefit is worth the price.

(2) Legally speaking, corporations must follow the laws of the states in which they operate. Philosophically speaking, this implies a legitimate and authoritative relationship between sovereign political entities and individual business entities that are involved in particular economic activities in markets. By examining the differences between how companies operate in the US vs. a Communist nation, for example, it is visible that the varying political philosophies of a place significantly affect the corporations themselves, and their implicit philosophies.

(3) It is an assumption that human rights law is relevant to corporate behavior, though this philosophical commitment is followed to greater or lesser degrees depending on the context. Companies create unethical products and promote their use all the time, e.g., technology companies might create AI products that violate privacy by collecting too much data, chatbots that manipulate and lie to users, or self-driving cars that injure people. However, despite these realities of how corporations might engage human rights, they still often at least claim to consider human rights.

(4) The field of business ethics gets into details on the right way to conduct business. Business ethics is a well-developed field of applied ethics, and is standard in business schools. However, theory and practice are, as with human rights (or just about anything), seldom completely aligned, and business practice often varies from theory.

Corporations often reduce their ethical requirements to the merest legal compliance. Commitments to social benefit are forgotten and human rights and business ethics are only considered insofar as they might cause bad press. When it comes to business, practical concerns often dominate the theoretical ideals that lie in their implicit philosophies.

However, a basic notion of social benefit out of self-interest, respect for law, respect for human rights, and basic business ethics, are a few philosophical ideas underpinning corporations. There are certainly many more. But these are sufficient to illustrate the point and begin the conversation.

4.2.2 The Catholic Church

The Catholic Church also embodies philosophies in its institution and practices, as well as in overt philosophical and theological commitments. More theoretically speaking, the purpose of the Catholic Church is to bring glory to God and advance the understanding and expression of the love of God around the world. One expression of the nature of this duty to share God's love with the world can be found in the *Catechism of the Catholic Church*, which concludes its prologue with: "The whole concern of doctrine and its teaching must be directed to the love that never ends. Whether something is proposed for belief, for hope, or for action, the love of our Lord must always be made accessible, so that anyone can see that all of the works of perfect Christian virtue spring from love and have no other objective than to arrive at love" (1995, p. 9–15). More practically-speaking, this involves evangelization, providing education and healthcare, pursuing justice and peace, the provision of sacraments to the faithful around the world, and so on. These pursuits are an expression of the Church's theological commitments to seeing every human being as a person made in the image and likeness of God, with intrinsic dignity and transcendent significance, as Jesus Christ embodied on Earth (Matthew 25: 31–46).

The practical results of the above theological commitments are a hierarchical system of parishes, dioceses, archdioceses, bishop's conferences, and finally the whole Catholic Church, providing a systematic approach to the distribution of sacraments, the spiritual and corporal works of mercy including education, healthcare, and so on. This rigid system also has numerous parallel systems in the form of orders of priests, brothers, sisters, monks, nuns, and others who take on specific tasks, such as missionaries who can go help anywhere they are needed.

This focus on glorifying God and protecting human dignity is quite a different objective than corporations, which are intrinsically economic in their *raison d'être*.

The Catholic Church is certainly involved in economic activities, but these economic activities exist for the ethical purposes dictated by theological and philosophical commitments. Corporations also tend to be much younger; Catholic Christianity has been focused on its ethical teachings for 2000 years (while certainly not living up to them much of the time, and even occasionally changing teachings (see, e.g., Curran, 2003; Noonan, 2005)).

4.3 What Are the Converged Upon Ethical Commitments?

4.3.1 The Corporate Perspective

Corporations and the Vatican, then, come from very different starting points when it comes to ethics. And yet, in the specific context of technology companies, they come to similar ethical principles. I will summarize the ethical commitments of four large tech companies, building upon Flahaux et al. (2023, p. 102–106), then compare these with some ideas from Catholic teaching.

4.3.1.1 Microsoft

Microsoft has six AI ethics principles:

1. Fairness: AI systems should treat all people fairly
2. Reliability & Safety: AI systems should perform reliably and safely
3. Privacy & Security: AI systems should be secure and respect privacy
4. Inclusiveness: AI systems should empower everyone and engage people
5. Transparency: AI systems should be understandable
6. Accountability: People should be accountable for AI systems (Microsoft, 2023)

As more practical applications of their principles, Microsoft also developed several ethical tools for product development: Judgment Call, the Envision AI Workshops, Impact Assessment, Community Jury, and a set of tools for ethically improving machine learning models (Green et al., 2021b).

4.3.1.2 IBM

IBM's principles are "the guiding values that distinguish IBM's approach to AI ethics":

1. The purpose of AI is to augment human intelligence
2. Data and insights belong to their creator
3. Technology must be transparent and explainable (IBM, 2023)

IBM also has five Pillars that are "our foundational properties for AI ethics":

1. Explainability: Good design does not sacrifice transparency in creating a seamless experience
2. Fairness: Properly calibrated, AI can assist humans in making fairer choices
3. Robustness: As systems are employed to make crucial decisions, AI must be secure and robust
4. Transparency: Transparency reinforces trust, and the best way to promote transparency is through disclosure
5. Privacy: AI systems must prioritize and safeguard consumers' privacy and data rights (IBM, 2023)

IBM operationalizes their pillars through several specialized AI toolkits focused on Explainability, Fairness, Adversarial Robustness, FactSheets, Privacy, Uncertainty Quantification, and Causal Inference (IBM, 2022; Green et al., 2021a).

4.3.1.3 Salesforce

Salesforce has core values and ethical use guiding principles. Their core values are:

1. Trust: We act as trusted advisors
2. Customer Success: When our customers succeed, we succeed
3. Innovation: We innovate together
4. Equality: Everyone deserves equal opportunities
5. Sustainability: We lead boldly to address the climate emergency (Salesforce, 2023)

These core values are a bit different from what either Microsoft or IBM describe in their principles and pillars. However, Salesforce's ethical use guiding principles are similar in some ways to Microsoft and IBM:

1. Human Rights: We work to ensure the direct use of our technologies upholds equal and inalienable protections
2. Privacy: We push the frontier of privacy best practice in our product design to enable customers to protect individuals' data
3. Safety: We aim to protect humans from direct harm from the use of our technology
4. Honesty: We oppose the use of our technology to knowingly spread disinformation or conspiracy theories
5. Inclusion: We create opportunity through equal access to technology (Salesforce, 2022)

Salesforce implements their values and guiding principles through their corporate strategy and accountability mechanisms (Green et al., 2022). In this way employees are incentivized to implement and participate in the values and principles.

4.3.1.4 Google

Google has seven AI Principles:

1. Be socially beneficial
2. Avoid creating or reinforcing unfair bias
3. Be built and tested for safety
4. Be accountable to people
5. Incorporate privacy design principles
6. Uphold high standards of scientific excellence
7. Be made available for uses that accord with these principles (Google, 2023)

Google also has four AI applications it will not pursue:

1. Technologies that cause or are likely to cause overall harm. Where there is a material risk of harm, we will proceed only where we believe that the benefits substantially outweigh the risks and will incorporate appropriate safety constraints.
2. Weapons or other technologies whose principal purpose or implementation is to cause or directly facilitate injury to people
3. Technologies that gather or use information for surveillance violating internationally accepted norms
4. Technologies whose purpose contravenes widely accepted principles of international law and human rights (Google, 2023)

For implementation, Google has categorized "general recommended practices for AI, fairness, interpretability, privacy, and safety," thus adding another set of ethical concerns, each of which also has further specifications, which can be quite detailed (Google, 2022). Google has used materials developed by universities and other sources to shape their thinking when it comes to ethical AI, for example the *Ethics in Technology Practice* materials (Walker, 2018; Vallor et al., 2018).

Each company uses its own language to describe its ethical commitments, and these words often cover similar expanses of the ethical terrain, for example, honesty and trust. I will try to combine some near-synonyms and show which themes emerge.

All four of the companies include a principle or pillar regarding privacy. Three of them explicitly mention fairness, and the company that does not, Salesforce, covers similar but not identical territory with the words equality and inclusion. Three companies explicitly mention safety; only IBM does not, but covers similar territory with "robustness." Accountability is explicitly mentioned by Google and Microsoft. Honesty is explicitly mentioned by Salesforce, while trust is mentioned by Salesforce and IBM, but for IBM, within the pillar of transparency. Transparency and explainability are most emphasized by IBM, and also mentioned by Microsoft, but not the other companies. Inclusion is explicitly mentioned by Salesforce and Microsoft. Human rights are explicitly mentioned by Google and Salesforce.

Other principles that are less commonly mentioned (mentioned by one company but not the others) are: (1) IBM: the purpose of AI is to augment human intelligence, and customers own their own data, (2) Salesforce: Customer Success, Innovation, and Sustainability, (3) Google: be socially beneficial, uphold scientific excellence, be available for use in accord with these principles, avoid harm (with proportionality in mind), do not create injurious weapons, and avoid norm-violating surveillance. Microsoft has no principles that are not also mentioned by the other companies.

Corporations, then, seem to have a good collection of moral norms. What does the Vatican have to say in reply?

4.3.2 *The Religious Perspective*

From the Catholic perspective there is a quick approach to AI ethics and a long approach. The Rome Call for AI ethics represents the quick, and Catholic social teaching the long.

4.3.2.1 The Rome Call for AI Ethics

One segment of the Vatican hierarchy of the Catholic Church has also put forward some norms in relation to AI. The Rome Call for AI ethics, a partnership of the Pontifical Academy for Life, Microsoft, and IBM, has six ethical principles:

1. Transparency: in principle, AI systems must be explainable
2. Inclusion: the needs of all human beings must be taken into consideration so that everyone can benefit and all individuals can be offered the best possible conditions to express themselves and develop
3. Responsibility: those who design and deploy the use of AI must proceed with responsibility and transparency
4. Impartiality: do not create or act according to bias, thus safeguarding fairness and human dignity
5. Reliability: AI systems must be able to work reliably
6. Security and privacy: AI systems must work securely and respect the privacy of users (RenAIssance Foundation, 2020)

These principles actually match Microsoft's to a remarkable degree, essentially aligning 1:1 but with a few words changed and the order shuffled (Table 4.1).

Since the Rome Call was specifically designed for bridging the gap from Church to company, this alignment should be no surprise, and also, might not serve as the best demonstration of the Catholic Church formulating its own norms in relation to AI. But it is remarkable in that the Church certainly has no problems with the principles, even if they might not be the fullest form of AI ethics that the Church can

Table 4.1 Description

Rome Call	Microsoft
1. Transparency	5. Transparency
2. Inclusion	4. Inclusiveness
3. Responsibility	6. Accountability
4. Impartiality	1. Fairness
5. Reliability	2. Reliability & Safety
6. Security and privacy	3. Privacy & Security

produce. What then, might be some components of a fuller version of a Catholic AI ethic?

4.3.2.2 Church Teachings Related to AI Ethics

Examining Church documents can provide deeper insights into what the Church teaches regarding the ethics of technologies. Most of the documents considered here are from the Church's social teaching, which can be helpfully reinterpreted as the Church's response to *social changes caused by technologies.*

Pope Leo XIII's encyclical letter *Rerum Novarum* in 1891 provides a starting place for Catholic interest in the effects of technologies on society. With the Industrial Revolution transforming the social world, this encyclical launched the Church's social teaching, saying: "The elements of the conflict now raging are unmistakable, in the vast expansion of industrial pursuits and the marvelous discoveries of science..." (1891, para. 1). Given the "misery and wretchedness" afflicting so many, the Church had to speak out (1891, para. 3). However, *RN* only uses the word "science" once, and the word technology not at all, indicating the nascency of these teachings.

Pope Pius XI's *Quadragesimo Anno* in 1931 begins to discuss "technical and economic progress" (1931, para. 72) for the first time, while at the same time being clear that its competency is in faith and morals, not "technique" (1931, para. 41). But this division of competencies does not excuse the Church from concerns for society: "...the Church... can in no wise renounce the duty God entrusted to her to interpose her authority... in all things that are connected with the moral law" (1931, para. 41). The Church must be concerned about the social impacts of technology, while not being an expert in the technology itself.

Pope John the XXIII's *Mater et Magistra* in 1961 begins to explicitly mention particular technologies in detail, and describes the Church's long-standing interest in the uses of technology for the sake of promoting the common good (1961, para. 37–47). The encyclical also makes clear that *Rerum Novarum* and *Quadragesimo Anno* were both about technology, even though they were less explicit about that specific word.

Pope John XXIII's *Pacem in Terris* in 1963 remains focused on technology, it being the explicit topic of two of the first three paragraphs of the encyclical. John

XXIII also specifically mentions Christians working in technology, emphasizing the need to keep science, technology, ethics, and wisdom as one whole (1963, para. 150). The encyclical also calls for nuclear disarmament and the banning of nuclear weapons (1963, para. 112).

The Second Vatican Council document *Gaudium et Spes*, 1965, considers the importance of examining science and technology, and appreciating their benefits, while remembering that they need to be directed towards good ends (Paul VI and Second Vatican Council, 1965, para. 57, etc.). This again reinforces the idea that technology is subject to ethical discernment. Technology does not justify itself; it is subject to evaluation and can be rejected if it is insufficiently ethical (Green, 2017, 2020). The Church affirms that *technological progress is not only good, it should be promoted* (1965, para. 64), and furthermore that science, technology, theology, and ethics ought to be integrated so that "religious culture and morality may keep pace with scientific knowledge and with the constantly progressing technology" (1965, para. 62).

Pope Paul VI's *Populorum Progressio* in 1967 clearly supports the development of the world and its peoples, including scientific, technological, and economic development. However, he reminds that this development is not an end in itself—it is always to be judged relative to human beings and human dignity, saying, "Economics and technology are meaningless if they do not benefit [the human person], for it is he they are to serve" (1967, para. 34).

Pope Paul VI's *Octogesima Adveniens*, 1971, increases the emphasis on scientific and technological progress found in recent encyclicals (1971, para. 22), while hope for the future becomes a new topic of concerns, saying that in the rush for technological development we risk the possibility of peace (1971, para. 19). It also asks a new question about the ideological nature of technology, saying that "universalized technology" is becoming "the overwhelming pattern of existence… without the question of its meaning being really asked" (1971, para. 29). The prophetic nature of this point is apparent: since the 1970s numerous technologically-oriented ideologies have sprung forth, such as transhumanism (Gouw et al., 2022), religions devoted to AI (Harris, 2017), and the growth of the "the reign of technology": the technocratic paradigm (Paul VI, 1971, para. 34).

Pope John Paul II's *Laborem Exercens* in 1981 was a deep investigation into the role of work in human nature, existence, and society. In his typically philosophical fashion, he dives into technology as a philosophical issue for labor, saying that technology is "a basic coefficient of economic progress; but" also raises questions that are "particularly charged with *content and tension… of an ethical and social character,*" thus constituting "a challenge for the Church" (1981, para. 5, emphasis original). This again affirms the Church's interest in technology, but subordinated to ethics.

Pope John Paul II's *Centisimus Annus* in 1991 spends much of its text considering the changed economic order of the time. However, it does raise the topic of ownership of intellectual property and technical skill, saying, "There exists another form of ownership which is becoming no less important than land: *the possession of know-how, technology and skill.* The wealth of the industrialized nations is based much more on this kind of ownership than on natural resources" (1991, para. 32). This realization is important for education, a major work of the Church.

Pope Benedict XVI's *Caritas in Veritate* in 2009 was significant for including a chapter just on technology. Benedict gathered many ideas from the tradition, for example, "The process of globalization could replace ideologies with technology, allowing the latter to become an ideological power that threatens to confine us… Were that to happen, we would all know, evaluate and make decisions about our life situations from within a technocratic cultural perspective to which we would belong structurally, without ever being able to discover a meaning that is not of our own making" (Benedict XVI, 2009, para. 70). In other words, we would be inescapably trapped in a web of our own ideology. Benedict XVI concludes with a defense of human freedom and a call for ethics: "*But human freedom is authentic only when it responds to the fascination of technology with decisions that are the fruit of moral responsibility.* Hence the pressing need for formation in an ethically responsible use of technology" (2009, para. 70, emphasis original). The phrase "responsible use of technology" is remarkable here, significantly preceding the contemporary secular and corporate conversation where the phrase has become commonplace (e.g. Quintanilla et al., 2019).

Pope Francis's *Laudato Si'*, 2015, is a landmark in Catholic teaching on technology, mentioning words with "techn-" roots 95 times, in 50 of its 246 paragraphs, thus demonstrating a growing interest in technology. Indeed, *Laudato Si*'s "Chapter 3: The Human Roots of the Ecological Crisis" can perhaps best be described as philosophy and theology of technology, and has a very significant discussion the importance of technology and its ethics (2015, para. 101–136). In particular, *Laudato Si'* para. 106 critiques the development of technology in a manner aligned with the technocratic paradigm, where the world is reduced to mere matter for manipulation, with no moral value or dignity. Contra technocracy, the Pope not only promotes the idea of human rights, but also of the moral value of nature and sustainability.

Pope Francis's *Fratelli Tutti*, 2020, again deeply engages technology and its impacts on the world, for example, the creation of a "massified" global culture while local communities are destroyed (2020a, para. 12), and says, "How wonderful would it be, even as we discover faraway planets, to rediscover the needs of the brothers and sisters who orbit around us" (2020a, para. 31). These statements display the value of diversity, inclusion, and caring for each other.

Several of Pope Francis's allocutions and a prayer intention are also worth mentioning here, as they particularly relate to AI. In his allocution to the participants in the "Common Good in the Digital Age" conference on September 27, 2019, Francis expressed his particular concern that the benefits of technological progress might not be shared with all people in the world, and that AI could be used to manipulate public opinion and put people out of work, driving a terrible inequality (Francis, 2019).

Pope Francis's prayer intentions for 2020 included a month of prayers for AI in November, saying "Artificial Intelligence: We pray that the progress of robotics and artificial intelligence may always serve humankind." This demonstrates not only the hope that AI will serve humankind, but also that God stands above all that happens on Earth, including AI (Daily Prayers, 2019; Vincent, 2020).

In February 2020 Cardinal Paglia delivered a message from Pope Francis to the first meeting of the Rome Call for AI Ethics. The speech aimed to affirm the gathering in its goals, saying "artificial intelligence, is at the very heart of the epochal change we are experiencing" and "We are beginning to glimpse a new discipline that we might call 'the ethical development of algorithms' or more simply 'algor-ethics'. This would have as its aim ensuring a competent and shared review of the processes by which we integrate relationships between human beings and today's technology." This statement supports the overall evaluation of technology by ethics as well as developing ways to properly integrate technology into society (Francis, 2020b).

In January 10, 2023, Pope Francis spoke again to the organizers of the Rome Call, saying "The Rome Call can be a useful tool for a common dialogue among all, in order to foster a humane development of new technologies." This call for dialogue has been a hallmark of Pope Francis's papacy (Francis, 2023).

Some of the ethical ideals that can be drawn from these, are, to summarize: valuing the common good and social benefit; the priority of ethics over technology; the importance of human dignity and human rights; the moral value of nature and sustainability; the value of diversity and inclusion; the importance of caring for our fellow human beings; and many more. These are notably similar to the ethical ideals put for by the four technology companies, however, the Catholic tradition differs from corporations in the depth with which these concepts are investigated and integrated with the tradition, which goes back thousands of years. The tech company principles look shallow by contrast.

Yet the question remains: why this convergence?

4.4 What Explains the Convergence?

It is central to the Catholic faith that God takes an interest in the affairs of Earth, and Christians throughout history have well-understood that their actions have temporal, and eternal, relevance. That some corporations would likewise take an ethical interest in AI—beyond what is legally required of them—might seem more surprising, however. But they do. What explains this convergence? There are at least three possible explanations.

First, the "simple, cynical" explanation: the convergence of secular corporate practice with traditional religious ethical values regarding AI principles might be a public relations move. These organizations are merely wishing to keep up appearances and look like good social institutions, while getting away with as much exploitation as they can behind the scenes. In this case, ethics work is only a cover, and taking these values from popular historical sources just means that the values are themselves popular and likely to resemble the cultural values of their locale.

Second, the "shared culture" explanation: tech corporations and the Catholic Church have come to similar formulations of AI ethics because they have come from the same culture in the first place. However, there are many more organizations which claim to have these shared values than just these few tech corporations and Catholicism. The Beijing ethics principles look just like Western ethical principles (International Research Center for AI Ethics and Governance, 2019–2023). And the UN Universal Declaration of Human Rights—as well as an extensive body of human rights laws—are accepted by most nations on Earth (United Nations, 1948). So if this explanation is correct, then most nations of the world seem to share the same cultural values.

Third, the "this is what works" explanation: any ethic that works at a broad scale and long term would need to be carefully delineated by the interaction of humans and their social and natural environment. Humans are capable of a multitude of behaviors, and many of them are likely to lead to unpleasant outcomes unless something stops them. Those influences which limit human behavior to exclude the most discordant actions and include the most concordant ones might become ethics, laws, and other social expectations. This explanation is not necessarily at odds with the "shared culture" explanation—perhaps the reason the culture is shared is because it works.

As an explanation, "this behavior works" would need to be sustainable over the long term and across a wide variety of social and environmental contexts. It would be, at least relatively, non-discordant with both individual human psychology, the needs of human society and civilization, and the environment. It would need to include very basic ideas such as safety, justice and fairness, accountability, and inclusiveness, among other things.

While a mere social explanation could be sufficient here, looking deeper into the ethical traditions of the West we might find resonances in natural law ethics. Rooted in the Greek and Roman philosophical traditions, integral to Catholic ethics, translated in secular human rights language via the medieval to early-modern "natural rights" discourse (Tierney, 1997), and still resonating in the naturalistic ethics discourse in contemporary philosophy (which is often actively anti-religious, e.g., Harris, 2010), natural law ethics roots ethical norms into human nature. And because all humans share our human nature, we can expect some ethical similarities across all human societies.

While corporations, in their commitments to human rights and other universalistic ethical principles, are not exactly doing natural law ethics, they are at least partially committing themselves to the philosophical descendants of natural law ethics. It might well be unreflective and based merely on intuition or popularity, but just as corporations are unreflectively philosophical in the other ways, the philosophy is there—real but not understood.

How things are is one thing, but how things ought to be is another thing entirely. Is it good that secular corporations, the Catholic Church, and many other organizations are mainly in agreement on many fundamental ethical premises that relate to AI? Perhaps these groups have all made the same error. Furthermore, with a preponderance of practitioners on one side and non-practitioners on the other, we might wonder if these ethical ideals, while using the same words, might actually mean different things due to the differences found in actual practice vs. more theoretical commitments.

The solution to the above wonderings is dialogue between these varying groups: an ongoing and broad-spectrum conversation including not just the voices of corporate technologists and Catholic academics, but of all people in the world affected by AI—which is everyone.

4.5 What *Should* We Do?

The point of this paper is not only to indicate the mere fact of convergence. The point of this paper is, based on this implicit ethical convergence, to determine what to do to promote social flourishing while surrounded by unprecedented technological change. The first step should be to continue to deeper "encounter" (to use one of Pope Francis's favorite words). So we should consider some effective means of encounter.

The below is not intended to be a fully comprehensive list, although I do include as many examples as I am aware of and potential opportunities that I can envision (but which I am not aware of actually existing). These practices are designed to facilitate the flow of information back and forth between very different social institutions, and may not be transferable to all other situations. These are offered here as examples to be considered, emulated, critiqued, dismissed, and so on, as the reader sees fit.

4.5.1 What Are Some Strategies for the Religious Side?

4.5.1.1 General Strategies Already in Practice

Various organizations have already adopted the following strategies for facilitating ethical dialogue on AI. For each strategy I present exemplars who are already doing this and the benefits and drawbacks of this particular mode of conversation.

(1) *Dialogue with Religious Believers in the Technology Industry*
Religious organizations can go to the religious faithful already in technology companies for technical guidance on technology ethics issues, and in return provide them ethical guidance for their roles in their companies.

Models: The Institute for Technology, Ethics, and Culture (ITEC, 2023); AI and Faith (2023).

Benefits: Technology professionals know the technology and its risks, and due to their religious faith are likely more motivated to help.

Drawbacks: Technology professionals often do not have an academic-level understanding of their religion, might be unwilling to challenge religious authority figures, might be unwilling or unable to translate their religious ethics in the secular language necessary for corporations to understand or utilize, and so on.

(2) *Dialogue with Non-religious Technology Leaders*

Religious organizations can go to non-believers who are prominent technologists or technology industry leaders and ask for their guidance.

Models: The Vatican's Minerva Dialogues demonstrate this strategy, engaging tech industry luminaries with the hope of gaining their insights and aiding their future thinking (Green, 2022; Lubov, 2023).

Benefits: These tech industry leaders know the technology and the industry as a whole. They are likely to be more influential than typical religious folks in tech companies (though not always). As non-believers they are typically willing to tell religious leaders what they really think, which is important since religious leaders can sometimes be isolated in echo chambers.

Drawbacks: These dialogues require trust, which takes time to build. Non-believing technologists will not necessarily know much about the religious side of things, nor really care. Their involvement totally depends on their level of motivation, which may easily falter due to lack of religious ties. Their involvement may be more for amusement or the altruistic aim of social benefit, so if amusement or prospective positive impact wanes, attention may falter and the dialogue dissipate.

(3) *Dialogue with Scholars of Religion*

Religious organizations can go to academics (believers or not) who know the faith and its resources, and get them to examine the technology from the direction of social impact, ethics, politics, theology, and philosophy, etc.

Models: Vatican's Dicastery for Culture and Education (DCE) AI Research Group, DCE East Asia AI discussion group, etc.

Benefits: Professional scholars really know the religious resources. Since they likely care deeply about the subject matter, they are likely to be motivated whether they are religious believers or not. They are also likely to be honest with religious leadership (since academics tend to be outspoken).

Drawbacks: Academics are sheltered in academia with limited connection to technological and pastoral reality. They may leave if they feel they are not deriving benefit (e.g., publications). There is a general academic bias against practical endeavors and towards more theoretical ones: big ideas to discuss, not big actions to do.

(4) *Faiths Encourage Technology Workers in General and Ethics Professionals in Particular*

Some of the folks working at technology companies are religiously motivated and they should be encouraged in their motivation to help the company's ethics efforts.

This should apply even more so to technology company employees who are doing ethics at these companies.

Models: Occasional messages from the Popes or other religious leaders to science and technology workers (Francis, 2019, 2022).
Benefits: encouragement of religious practitioners within corporate technology settings
Drawbacks: The general approach might generate small amounts of encouragement over a broad, shallow, population of workers, thus not being very effective.

(5) *Dialogue with Secular Multistakeholder Organizations*
Religious organizations can be involved (or become involved) in organizations and conferences that are interested in technology questions, like UNESCO, the ITU, SXSW, etc. These present networking opportunities and also help keep the religious organization's eyes open and communication channels developing and/or open.

Models: The Vatican's communications office and its representatives (Green, 2022).
Benefits: The opportunity to gather information from broad and varying sources, as well as to dispel popular myths about religion, science, and technology.
Drawbacks: This requires time, money, preparation, travel, networking, follow up, etc.

(6) *Dialogue on Science and Religion*
Rather than aim directly at discussion of technology ethics, religious organizations can aim more at the foundations underlying the possibility of technology ethics dialogues. The science and religion dialogue can open the door to the technology and ethics dialogue. This provides a chance to demystify science, emphasize Christians who are or were scientists and technologists, quash fear-mongering and rejection of science and technology, and develop a balanced approach to science and technology based on social impact as judged by that religion's leaders and people.

Models: The Center for Theology and the Natural Sciences (CTNS) at the Graduate Theological Union in Berkeley, as well as the Zygon Center for Religion and Science at the Lutheran School of Theology at Chicago are institutions dedicated to science and religion dialogue, and have already expanded their scope to include technology and ethics (CTNS, 2023; Zygon, 2023)
Benefits: The important theoretical outcome of this approach is to dispel myths and promote seeing truth as a unity: all truth is one; no contradictions; science, theology, technology and ethics all cohere. This also promotes the practical outcome of encouraging religious faithful to support science and technology while not abandoning their faith, thus allowing religion and science to engage in closer alignment.
Drawbacks: These efforts require time, labor, money, and interest.

(7) *Stronger Teaching Emphasis on Technology, Religion, and Ethics*
Schools and universities, particularly religious ones, can work harder to think about the social impacts of technology. This should involve research, education of

students and the public, targeted education at lawmakers and other people of influence, work with technology companies, and pastoral education.

Models: Most Catholic engineering schools have an ethics component to their curriculum, but they vary widely in intensity and scope. Other religious and secular schools may also have ethics emphases.

Benefits: Educational institutions already exist and already contain much of the talent necessary for making efforts like these successful. Getting stronger foundations for the religion, ethics, science, and technology dialogues can be potentially culture-shifting and world-changing.

Drawbacks: Academics are unlikely to do things like this unless it is incentivized: they need to be rewarded for doing things that do not result in traditional academic rewards (i.e., tenure and promotion).

4.5.1.2 Potential Strategies Not Currently in Extensive Practice

(1) *Dialogue on Technology and Ethics in Churches*

In this strategy churches would host speakers or other events to talk about contemporary issues in technology and ethics in a coordinated effort to push forward this dialogue.

Models: certainly churches have speakers, but a concerted effort at a continuing series of talks on AI would be unusual

Benefits: Allows for a grass-roots development of interest, knowledge, and talent among church-goers; potential to gather dormant talented individuals to the cause.

Drawbacks: requires money and time, hard to scale, potentially low impact, would require follow-up in order to be most effective, needs an organizational structure to respond to any call-to-action that might accompany the dialogues

(2) *Practical Technology Education in Churches*

Like the above dialogic approach, but with a practical focus: turn churches into significant local centers for technology learning (Marzani, 2016).

Models: churches with classes on computer use, skills development, etc., are similar.

Benefits: directly prepares people in a religious setting (whether believers or not) for potential careers in technology, thus providing a pipeline into the tech industry and helping people get better jobs.

Drawbacks: This would be a heavy lift to get technologists into churches teaching technology and engineering to people of highly variable backgrounds.

(3) *Church Collaboration with Technology Companies*

More direct partnerships with tech companies to visualize new possibilities and update current technology systems in religions and open religious groups' eyes to the reality of coming technology.

Models: unknown

Benefits: Not only would this help to make churches more technologically up-to-date, it would also incentivize direct connections to technology providers which are mutually beneficial.

Drawbacks: Without some charity or donations-in-kind from the tech companies this will likely not occur without another major source of money, and money for maintenance.

4.5.2 What Are Some Strategies for the Corporate Side?

4.5.2.1 General Strategies Already in Practice

(1) *Promote Ethics through Faith-Based Employee Resource Groups*

Many companies are forming employee resource groups (ERGs) based on various identity traits, one of which is religion or faith. These groups could collaborate with corporate ethics efforts in order to facilitate development of more ethically-oriented corporate cultures.

Models: various ERGs and an organization supporting them: the Religious Freedom and Business Foundation (RFBF) (Religious Freedom and Business Foundation, 2023).

Benefits: promote community at work, and companies can promote ethical initiatives within the company.

Drawbacks: These groups are likely to be a minority of the company in many cases, or the particularism of religions might not integrate with the overall corporate goals, or ethics might come to be associated only with religion within the company.

(2) *Corporate Chaplains Serving at Technology Companies*

Some technology companies are using corporate chaplains—pastors with an interest in helping believers at companies—to help them maintain motivation, morale, and direction in their companies.

Models: Major technology companies; one network of these chaplains is Corporate Chaplains of America (Corporate Chaplains of America, 2023).

Benefits: Religious believers in technology companies obtain desired counseling services and guidance which improves morale and productivity.

Drawbacks: Most chaplains require monetary compensation and some may question the cost-to-benefit value. There may be a fear of looking sectarian even if there are chaplains of every faith. Corporations may fear looking not-serious or anti-intellectual by facilitating employee religiosity.

(3) *Corporate Engagement of Religious Communities Regarding Ethical Product Development*

While some technology companies are starting to talk to communities affected by the technologies they are developing, these tend to be secular or activist-focused.

Models: one example would be the "Community Jury" tool for ethical discussion and guidance as described by Microsoft (Microsoft, 2022).

Benefits: by including religious communities in technology discussions, both sides will gain important information and a broader perspective on new technologies which they might not otherwise get.

Drawbacks: cost in terms of money and effort, and any community engagement is necessarily always limited.

(4) *Corporate Leadership Seeking Guidance from Religious Leadership*

In this strategy, corporate leaders get periodic guidance from religious and moral leaders on particular issues. This is more than mere chaplaincy, which can often have a more personal focus; this is seeking advice on major corporate decisions. These occasions are not much discussed, but they do happen.

Models: corporate leaders turn to their pastors, rabbis, imams, gurus, etc., for guidance in making corporate decisions.

Benefits: typically, these relationships already exist and just need to be called upon. Consulting more sources often results in better decision-making, especially since executives are specialized in sorting information from various sources and acting on them.

Drawbacks: it can be scandalous to be "outed" as seeking corporate guidance from non-corporate leaders. As an extreme example, the revelation that U.S. President Ronald Reagan's wife Nancy was consulting astrologers about important government business was a shock to many people (Roberts, 1988).

(5) *Dialogue with Multistakeholder Organizations that Include Religious Organizations*

Create network organizations that include corporate technology producers and civil society in order to develop means of communication between the creators of technologies and those affected by their products.

Models: the Partnership on AI, which includes many corporations as well as groups from civil society, including two religious groups: OPTIC and AI & Faith (Partnership on AI, 2023; International Dominican Foundation, 2022; AI & Faith, 2023).

Benefits: improved understanding on both corporate and religious sides, hopefully leading to improved decision-making and improved results.

Drawbacks: maintaining networks and their coordination is costly in time, money, and effort.

4.5.2.2 Potential Strategies Not Currently in Extensive Practice

(1) *Direct Appeal for Dialogue and Action from Corporations to Religions*

Corporations could talk directly and overtly to religious groups to get their input on technological impacts.

Models: perhaps no exact models; the Rome Call might be the closest, though it is not precisely this idea (RenAIssance Foundation, 2020).

Benefits: direct, open conversations remove stigma from previously hidden conversations. Due to their openness, these conversations might be honest—or the reverse, due to watching eyes.

Drawbacks: there is a potential for stigma for both religion and corporations. Religions for being too interested in money and temporal affairs, corporations for being too interested in things which are not money-oriented and temporal enough.

(2) *Corporate Promotion of Scientific and Technological Education in Churches*
Corporations could talk directly to religious groups to promote scientific and technological education (essentially "evangelism" of science and technology to churches)

Models: Many tech companies promote science and technology education in schools; this would be an extension to a new set of religiously-oriented locations. Similar efforts by the American Association for the Advancement of Science's (AAAS) Dialogue on Science, Ethics, and Religion (DoSER) exist, such as "Scientists in Synagogues" and "Science for Seminaries" (AAAS, 2023; Sinai and Synapses, 2023; Science for Seminaries, 2023).
Benefits: Many more people would gain important information and access about science and technology. It would reduce fear due to unfamiliarity, and build trust.
Drawbacks: This would be costly and involve a lot of effort. Additionally, there would be a strangeness of it due to it seeming to be "not our job" or "someone else's problem".

(3) *Corporations Amplifying Positive Religious Messages on Technology*
Building upon X.5.1.1. idea 4) above, when a religious leader makes a significant positive statement on technology, corporations can amplify that message to their employees or others.

Models: unknown, though perhaps various social media channels and comment boxes might be analogous.
Benefits: promotion of ethics or other interesting ideas relating to technology.
Drawbacks: may create a hostile environment if sectarianism intrudes, and the appearance of bias if some religions are more represented than others, which is likely, given, e.g., Pope Francis's particular interest in AI.

Overall, there are many means of communication between religions and corporations that could benefit both sides. But the obstacles are real and the impacts might seem abstract or distant, thus reducing motivation.

4.6 Conclusion

As described above, a lot of these measures could be construed as "corporations doing philosophy" and therefore can only be viewed with suspicion, like Otto in "A Fish Called Wanda" (who believed Aristotle was Belgian). But while corporations should certainly be careful when doing philosophy, doing philosophy is, in fact, unavoidable. In a very practical sense, philosophies are lived and merely by existing corporations silently articulate philosophy. So they might as well get better at it, by talking to those who know something about the subject; and while academic philosophers might seem to be the first choice for such endeavors, diverse religious scholars might well be the more practical conversation partners.

There are many technology company initiatives aimed at supporting traditional education whether through schools or universities (and not only technology companies; lots of companies promote education because not only do they need future workers, but it's also good public relations), perhaps it is also time for tech companies to support non-traditional places for education, such as churches, synagogues, mosques, and other religious meeting places. These locations also do not only need to be in the developed West—less-developed nations would likely be prime locations for educational work, with religious institutions as "delivery" points.

At the same times, religious organizations should promote ethics in technology companies, not merely as external critics, but as collaborators in trying to heal the world. There are many religious people inside of tech companies who would probably like to help with such initiatives.

This is a heavy lift on both sides, but without it we risk an ever-growing rift between human technological power on one side and traditional sources of meaning and wisdom on the other.

To return to C.P. Snow's *Two Cultures* problem, what the world needs now are synthetic specialists: people who understand multiple disparate areas of human endeavor and can hold them together and integrate them and share that understanding of integration with the world, not just literature and physics, as in Snow's day, and not just science and religion as in the more theoretical objectives of recent decades. Theory is vitally important, yes, but it can also become a luxury when the world is teetering near disaster. The time for luxury is passing into a time of practical emergency (Green, 2023, p. 176). We need to get ethics and technology under control right now, and that will require vastly more people working on it, and perhaps more people working on a particular ethical problem—AI—than have ever worked on any one specific ethical problem ever before.

To quote Nick Bostrom, this is "philosophy with a deadline" (2014, p. 314). And if we get it wrong, the deadline might actually mean dead. It is one thing, as with automation in the past, to externalize human muscle and have it applied back to us, as in the Industrial Revolution. It is another thing to externalize human intelligence and then have it applied back to us, as in the current AI revolution. We are already seeing the frightening disruptive capacities of this new technology turning the world upside down. If we want to have any hope of getting this technology under control, we need to work now as if our lives depend on it.

References

AI & Faith. (2023). https://aiandfaith.org/

American Association for the Advancement of Science. (2023). *Dialogue on science, ethics, and religion*. AAAS. https://www.aaas.org/programs/dialogue-science-ethics-and-religion

Benedict XVI. (2009). Caritas in Veritate. *Vatican*. https://www.vatican.va/content/benedict-xvi/en/encyclicals/documents/hf_ben-xvi_enc_20090629_caritas-in-veritate.html

Bostrom, N. (2014). Superintelligence. Oxford University Press.

Catechism of the Catholic Church. (1995). Doubleday.

Center for Theology and the Natural Sciences. (2023). *Graduate theological union*. https://www.ctns.org/
Cleese, J., & Crichton, C. (1988). *A fish called Wanda*. Metro-Goldwyn-Mayer.
Corporate Chaplains of America. (2023). https://chaplain.org/
Curran, C. E. (2003). *Change in official catholic teachings* (Readings in Moral Theology no. 13). Paulist Press.
Daily Prayers. (2019). Pope Francis' 2020 Monthly Intentions. *Daily Prayers*. https://www.dailyprayers.org/popes-monthly-intentions/pope-francis-2020-monthly-intentions/
Flahaux, J. R., Green, B. P., & Skeet, A. G. (2023). *Ethics in the age of disruptive technologies: An operational roadmap (The ITEC handbook)*. Markkula Center for Applied Ethics. https://www.scu.edu/institute-for-technology-ethics-and-culture/
Francis. (2015). Laudato Si: On care for our common home. *Vatican*. https://www.vatican.va/content/francesco/en/encyclicals/documents/papa-francesco_20150524_enciclica-laudato-si.html
Francis. (2019). Address of his Holiness Pope Francis to the participants in the seminar 'The Common Good in the Digital Age,' Organized by the Dicastery for Promoting Integral Human Development (DPIHD) and the Pontifical Council for Culture (PCC). *Vatican*. https://www.vatican.va/content/francesco/en/speeches/2019/september/documents/papa-francesco_20190927_eradigitale.html
Francis. (2020a). Fratelli Tutti. *Vatican*. https://www.vatican.va/content/francesco/en/encyclicals/documents/papa-francesco_20201003_enciclica-fratelli-tutti.html
Francis. (2020b). Meeting with the participants in the plenary assembly of the pontifical academy for life: Address prepared by Pope Francis, Read by H.E. Archbishop Paglia, President of the Pontifical Academy for Life. *Vatican*. https://www.vatican.va/content/francesco/en/speeches/2020/february/documents/papa-francesco_20200228_accademia-perlavita.html
Francis. (2022). Address of his Holiness Pope Francis to the pontifical academy of science. *Vatican*. https://www.vatican.va/content/francesco/en/speeches/2022/september/documents/20220910-plenaria-pas.html
Francis. (2023). Address of His Holiness Pope Francis to participants in the 'Rome Call' meeting promoted by the RenAIssance Foundation. *Vatican*. https://www.vatican.va/content/francesco/en/speeches/2023/january/documents/20230110-incontro-romecall.html
Google AI. (2022). Responsible AI practices. *Google AI Blog*. https://ai.google/responsibilities/responsible-ai-practices/
Google AI. (2023). Artificial intelligence at google: Our principles. *Google AI Blog*. https://ai.google/principles/
Gouw, A. M., Green, B. P., & Peters, T. (2022). *Religious transhumanism and its critics*. Lexington.
Green, B. P. (2017). The catholic church and technological progress: Past, present, and future. *Religions*. https://www.mdpi.com/2077-1444/8/6/106
Green, B. P. (2020). *Ethics is more important than technology*. Markkula Center for Applied Ethics. https://www.scu.edu/ethics/all-about-ethics/ethics-is-more-important-than-technology/
Green, B. P. (2022). The Vatican and artificial intelligence: An interview with Bishop Paul Tighe. *Journal of Moral Theology, 11*(1). https://jmt.scholasticahq.com/article/34131-the-vatican-and-artificial-intelligence-an-interview-with-bishop-paul-tighe
Green, B. P. (2023). What theology and science (and Ethics and Technology) can learn from indigenous scholars. *Religious Studies Review*. https://onlinelibrary.wiley.com/doi/full/10.1111/rsr.16543
Green, B. P., Lim, D., & Ratté, E. (2021a). *Responsible use of technology: The IBM case study*. World Economic Forum & Markkula Center for Applied Ethics. https://www3.weforum.org/docs/WEF_Responsible_Use_of_Technology_The_IBM_Case_Study_2021.pdf
Green, B. P., Lim, D., & Ratté, E. (2021b). *Responsible use of technology: The microsoft case study*. World Economic Forum & Markkula Center for Applied Ethics. https://www.weforum.org/whitepapers/responsible-use-of-technology-the-microsoft-case-study
Green, B. P., Ratté, E., Smallwood, R., & Zhang, C. (2022). *Responsible use of technology: The salesforce case study*. World Economic Forum & Markkula Center for Applied Ethics.

https://www3.weforum.org/docs/WEF_Responsible_Use_of_Technology_Salesforce_Case_Study_2022.pdf

Harris, S. (2010). *The moral landscape*. The Free Press.

Harris, M. (2017). Inside the first church of artificial intelligence. *Wired Magazine*. https://www.wired.com/story/anthony-levandowski-artificial-intelligence-religion/

IBM. (2022). *Trusted AI*. IBM. https://research.ibm.com/teams/trusted-ai

IBM. (2023). *AI ethics*. IBM. https://www.ibm.com/artificial-intelligence/ethics

Institute for Technology Ethics, and Culture (ITEC). (2023). *Markkula Center for Applied Ethics*. https://www.scu.edu/institute-for-technology-ethics-and-culture/

International Dominican Foundation. (2022). *Order of preachers for technology, informatics & communication*. International Dominican Foundation. https://internationaldominicanfoundation.org/cause/international-research-network-optic-order-of-preachers-for-technology-informatics-and-communication/

International Research Center for AI Ethics and Governance. (2019–2023). *Beijing artificial intelligence principles*. International Research Center for AI Ethics and Governance. https://ai-ethics-and-governance.institute/beijing-artificial-intelligence-principles/

John Paul II. (1981). Laborem Exercens. *Vatican*. https://www.vatican.va/content/john-paul-ii/en/encyclicals/documents/hf_jp-ii_enc_14091981_laborem-exercens.html

John Paul II. (1991). Centisimus Annus. *Vatican*. https://www.vatican.va/content/john-paul-ii/en/encyclicals/documents/hf_jp-ii_enc_01051991_centesimus-annus.html

John XXIII. (1961). Mater et Magistra. *Vatican*. https://www.vatican.va/content/john-xxiii/en/encyclicals/documents/hf_j-xxiii_enc_15051961_mater.html

John XXIII. (1963). Pacem in terris: On establishing universal peace in truth, justice, charity, and liberty. *Vatican*. https://www.vatican.va/content/john-xxiii/en/encyclicals/documents/hf_j--xxiii_enc_11041963_pacem.html

Leo XIII. (1891). Rerum Novarum. *Vatican*. https://www.vatican.va/content/leo-xiii/en/encyclicals/documents/hf_l-xiii_enc_15051891_rerum-novarum.html

Lubov, D. C. (2023). Pope Francis urges ethical use of artificial intelligence. *Vatican News*. https://www.vaticannews.va/en/pope/news/2023-03/pope-francis-minerva-dialogues-technology-artificial-intelligenc.html

Marzani, S. (2016). *Personal communication*. Santa Clara University.

Microsoft. (2022). *Community jury*. Microsoft. https://learn.microsoft.com/en-us/azure/architecture/guide/responsible-innovation/community-jury/

Microsoft. (2023). *Our approach*. Microsoft. https://www.microsoft.com/en-us/ai/our-approach

Noonan, J. T., Jr. (2005). *A church that can and cannot change: The development of catholic moral teaching*. Notre Dame University Press.

Partnership on AI to Benefit People and Society. (2023). https://partnershiponai.org/

Paul VI. (1967). Populorum progressio: Encyclical of Pope Paul VI on the development of peoples. *Vatican*. https://www.vatican.va/content/paul-vi/en/encyclicals/documents/hf_p-vi_enc_26031967_populorum.html

Paul VI. (1971). *Octogesima Adveniens*. https://www.vatican.va/content/paul-vi/en/apost_letters/documents/hf_p-vi_apl_19710514_octogesima-adveniens.html

Paul VI and Second Vatican Council. (1965). Pastoral constitution on the Church in the Modern World: Gaudium et Spes. *Vatican*. https://www.vatican.va/archive/hist_councils/ii_vatican_council/documents/vat-ii_const_19651207_gaudium-et-spes_en.html

Pius XI. (1931). Quadragesimo Anno. *Vatican*. https://www.vatican.va/content/pius-xi/en/encyclicals/documents/hf_p-xi_enc_19310515_quadragesimo-anno.html

Quintanilla, P., Allison-Hope, D., Darnton, H., & Hunter, A. (2019). *Responsible use of technology*. World Economic Forum. https://www3.weforum.org/docs/WEF_Responsible_Use_of_Technology.pdf

Religious Freedom and Business Foundation. (2023). https://religiousfreedomandbusiness.org/

RenAIssance Foundation. (2020). The Rome call for AI ethics. *Vatican*. https://www.romecall.org/wp-content/uploads/2022/03/RomeCall_Paper_web.pdf

Roberts, S. V. (1988). White House confirms reagans follow astrology, up to a point. *The New York Times*. https://www.nytimes.com/1988/05/04/us/white-house-confirms-reagans-follow-astrology-up-to-a-point.html

Salesforce. (2022). Ethical use policy. *Salesforce*. https://www.salesforce.com/company/intentional-innovation/ethical-use-policy/

Salesforce. (2023). Our story. *Salesforce*. https://www.salesforce.com/company/our-story/

Science for Seminaries. (2023). *AAAS DoSER*. https://www.scienceforseminaries.org/

Scientists in Synagogues. (2023). *Sinai and synapses*. https://sinaiandsynapses.org/scientists-in-synagogues/

Smith, A. (1994). *An inquiry into the nature and causes of the wealth of nations*. The Modern Library.

Snow, C. P. (1998). *The two cultures*. Cambridge University Press.

Tierney, B. (1997). *The idea of natural rights*. Eerdmans.

United Nations. (1948). *Universal declaration of human rights*. United Nations General Assembly. https://www.un.org/en/about-us/universal-declaration-of-human-rights

Vallor, S., Green, B. P., & Raicu, I. (2018). *Ethics in technology practice*. Markkula Center for Applied Ethics. https://www.scu.edu/ethics-in-technology-practice/

Vincent, J. (2020). Pope Francis urges followers to pray that AI and robots 'always serve mankind'. *The Verge*. https://www.theverge.com/2020/11/11/21560076/pope-francis-ai-for-good-pray-serve-mankind

Walker, K. (2018). Google AI principles updates, six months in. *Google Blog*. https://www.blog.google/technology/ai/google-ai-principles-updates-six-months/

Zygon Center for Religion and Science. (2023). *Lutheran School of Theology at Chicago*. https://lstc.edu/academics/zygon/

Chapter 5
Rome Call for AI Ethics: Considerations for a Philippine Response

Robert James M. Boyles

Abstract This paper surveys the three impact areas identified in the "Rome Call for AI Ethics", focusing on how they relate to the Philippine context. Resulting from a conference organized by the Vatican's Pontifical Academy for Life, the said document advances an ethical approach to the development and employment of artificially intelligent systems. Specifically, it centers on three impact areas, namely: ethics, education, and rights. In this article, the current state of these areas in the Philippines are examined (i.e., in order to provide an initial, practicable backdrop for any prospective response). By treating the said impact areas as initial assessment indicators, this study argues that the Philippine government is not that primed to respond to the Vatican's call in light of its track record of handling concerns related to ethics, education, and rights. Additionally, this research also holds that, although there might be certain challenges in further reforming the standing of these impact areas in the Philippines, there are indications that things are moving towards a positive direction.

Keywords Artificial Intelligence Ethics · Artificial Intelligence Governance · Catholic Church · Rome Call for AI Ethics · Philippines

R. J. M. Boyles (✉)
Department of Philosophy, Southeast Asia Research Center and Hub, De La Salle University, Manila, Philippines
e-mail: robert.boyles@dlsu.edu.ph

L. Checketts, B. S. B. Chan (eds.), *Social and Ethical Considerations of AI in East Asia and Beyond*, Philosophy of Engineering and Technology 47,
https://doi.org/10.1007/978-3-031-77857-5_5

5.1 Introduction

Considering the present and prospective effects of artificial intelligence (AI), several organizations, states, and the like have already advocated for the proper development and use of these kinds of technologies. The United States, for instance, has enacted its National AI Initiative Act of 2020 into law, which seeks to sustain the country's global leadership in the production of trustworthy AIs (Roberts et al., 2021, 12). Correspondingly, the Organisation for Economic Co-operation and Development (OECD) released a document, entitled "Recommendation on Artificial Intelligence", which also promotes the responsible oversight of trustworthy AI systems (Yeung, 2020, 27). Yet another call for the advancement of ethical intelligent machines has come from the Holy See.

In 2020, the Vatican, through its Pontifical Academy for Life and non-profit arm RenAIssance Foundation, issued the "Rome Call for AI Ethics", a document that promotes an ethical approach towards the development and use of AIs. Among the initial signatories of this pronouncement include tech giants IBM and Microsoft, Italy's Ministry of Innovation, and so on (Lubov, 2023). The general motivation behind this call is to "promote a sense of shared responsibility among international organizations, governments, institutions and technology companies in an effort to create… artificial intelligence that serves every person and humanity as a whole" (Pontifical Academy for Life, 2020, 2). This thought is further elucidated in three impact areas, namely: (1) ethics, (2) education, and (3) rights.

The first impact area, ethics, highlights the idea that AIs should be conceptualized, modeled, and used to cater to the needs and interests of human beings, including their living environments (Pontifical Academy for Life, 2020, 5). On the other hand, the main aim of the impact area on education is to create curricula or programs that underline social inclusion- and individual respect-related opportunities aided by AIs, while also considering the potential critical issues that they might pose (8). The last domain, rights, encourages all concerned stakeholders to consciously reflect on, if not code in pertinent values, principles, and legal regulations, the idea that AI systems ought to be made and employed for humanity and the entire world (9).

Supposedly, if different institutions, governments, and technology companies consider and focus on the impact areas mentioned above, the notion that AIs would favorably affect the present and future would somehow be ensured. However, positively responding to pronouncements such as the "Rome Call for AI Ethics" has also posed significant challenges for some. For one, several governments, such as the Philippines, have by and large experienced issues in complying with the said types of public declarations.

The Philippines has had problems in responding to calls issued by the global community. For instance, some studies, like that of La Viña and Reyes (2021), have given critiques concerning the country's climate change initiatives in relation to the Paris Agreement. In another case, in terms of handling human rights concerns, the Philippine government has also been scrutinized for a number of its lapses (Pernia,

2019; Regilme, 2021), which sheds light on how the country has generally fared in relation to decrees such as the *Universal Declaration of Human Rights* (United Nations General Assembly, 1948).

In terms of responding to the "Rome Call for AI Ethics", perhaps a more reasonable approach for the Philippines would be to take things one step at a time. The present study, thus maintains that if the Philippine government was to actually respond to the call, it should be properly informed first about the local, extant conditions that relate with the three impact areas indicated in the document. So, in this paper, the present state of these domains in the Philippines are looked into with the modest goal of offering an initial, practicable backdrop for any prospective response.

Following the idea of treating certain conditions as preliminary assessment indicators to check a state's capability of addressing large-scale problems (Boyles et al., 2022), this study makes use of the impact areas stated in the "Rome Call for AI Ethics" as measures to demonstrate that the Philippine government is not that ready to respond to the Vatican's call. Such is the case in light of its track record of handling concerns related to ethics, education, and rights. Nevertheless, this chapter also claims that, although there might be some challenges in reforming the standing of these areas in the Philippines, there are a few indications that things are moving towards a more positive direction.

To argue that the Philippine government is not presently primed to respond to the "Rome Call for AI Ethics", the following section details the domains concerning ethics, education, and rights. The next section provides a brief background on how the Philippines has generally responded to the challenges ensuing from the introduction of AI systems. The subsequent section, then, examines the local context of the impact areas mentioned earlier. The primary goal of this section is to provide a more concrete, yet still precursory and non-exhaustive, backdrop for any future response, specifically by citing government undertakings that could somehow shed light on the level of importance that the Philippines has accorded these impact areas. For one, if certain government activities or initiatives in the past seem to exemplify that they run counter to the ideals espoused in the said call, then it may be contended that formulating any response would be quite a challenge. The final part of this article gives some concluding remarks and a few recommendations for prospective research.

5.2 Rome Call for AI Ethics

The "Rome Call for AI Ethics" resulted from a conference organized by the Vatican's Pontifical Academy for Life (Lubov, 2023). In general, the document advances an ethical approach to the development and employment of AI systems. On the one hand, it acknowledges that the latter kinds of technologies do bring forth positive outcomes, like how these could further improve the personal and social lives of humans (Pontifical Academy for Life, 2020, 3). The same document, however, also forewarns that any drastic changes stemming from AIs should serve the best interests of all, wherein no one should be left behind.

Furthermore, the call encourages that all concerned stakeholders work together, so that the needs of everyone, especially those affected by the digital transformations instigated by AI technologies, would be properly addressed (Pontifical Academy for Life, 2020, 3). Concurrently, it promotes the idea of having a shared responsibility with regard to the whole technological innovation process of AIs, including their design, distribution, and use (3). By striving to meet such ends, the same document maintains that the "values and principles that we are able to instill in AI[s] will help to establish a framework that regulates and acts as a point of reference for digital ethics, guiding our actions and promoting the use of technology to benefit humanity and the environment" (4).

So, per the "Rome Call for AI Ethics", an ethical approach towards the development and use of AI systems should be an end goal. The call further claims that

> ...[n]ow more than ever, we must guarantee an outlook in which AI is developed with a focus not on technology, but rather for the good of humanity and of the environment, of our common and shared home and of its human inhabitants, who are inextricably connected. In other words, a vision in which human beings and nature are at the heart of how digital innovation is developed, supported rather than gradually replaced by technologies that behave like rational actors but are in no way human. It is time to begin preparing for [a] more technological future in which machines will have a more important role in the lives of human beings, but also a future in which it is clear that technological progress affirms the brilliance of the human race and remains dependent on its ethical integrity. (Pontifical Academy for Life, 2020, 4)

Specifically, the ethical approach advocated for by the call underscores three impact areas, namely: (1) ethics, (2) education, and (3) rights.

5.2.1 Ethics

The first area highlighted by the "Rome Call for AI Ethics", ethics, centers on the view that AIs ought to be conceptualized, modeled, and mainly used to cater to the needs and interests of humans, including their living environments (Pontifical Academy for Life, 2020, 5). In one sense, this is predicated on the idea that everyone is born equal and free, specifically in terms of rights and dignity. The impact area on ethics also cites the *Universal Declaration of Human Rights* (United Nations General Assembly, 1948), which emphasizes that humans, guided by their conscience and capacity of reason, should treat each other in the spirit of fellowship (Pontifical Academy for Life, 2020, 5). Hence, it follows that the rights and freedoms of all individuals ought to be safeguarded whenever one makes use of AI systems, especially that discriminations related to the latter have already been previously raised (Zuiderveen Borgesius, 2020; Heinrichs, 2022).

Moreover, the area on ethics also accentuates the thought that the only time that technological advancements would yield true progress is if they meet three specific conditions (Pontifical Academy for Life, 2020, 6). These conditions are as follows:

> It must include every human being, discriminating against no one; it must have the good of humankind and the good of every human being at its heart; finally, it must be mindful of the complex reality of our ecosystem and be characterized by the way in which it cares for and protects the planet (our "common and shared home") with a highly sustainable approach, which also includes the use of artificial intelligence in ensuring sustainable food systems in the future. (Pontifical Academy for Life, 2020, 6)

The impact area on ethics further emphasizes that humans should always be made aware whenever they are interacting with intelligent systems. This ties in with recent observations that many people are not that informed of when they are already interacting with AIs (Hyken, 2017).

Finally, this impact area focuses on giving importance to the most vulnerable (Pontifical Academy for Life, 2020, 6). It provides a warning that AI systems should not be used to exploit or take advantage of the said individuals. Rather, advanced technologies must be employed to aid human beings, empowering and enabling them, and they should also be used to take care of the planet.

5.2.2 *Education*

The main aim of the impact area on education, on the other hand, is to create curricula or programs that underscore social inclusion- and individual respect-related opportunities aided by AI systems, while also considering the potential critical issues they might pose (Pontifical Academy for Life, 2020, 8). This impact area further states that:

> Transforming the world through the innovation of AI means undertaking to build a future for and with younger generations. This undertaking must be reflected in a commitment to education, developing specific curricula that span different disciplines in the humanities, science and technology, and taking responsibility for educating younger generations. This commitment means working to improve the quality of education that young people receive; this must be delivered via methods that are accessible to all, that do not discriminate and that can offer equality of opportunity and treatment. (Pontifical Academy for Life, 2020, 7)

In addition, this particular area advances the democratization of education for all human beings, guided by the principles of fairness and solidarity.

In order to attain the aim of democratizing access to education, the "Rome Call for AI Ethics" also promotes and supports lifelong learning (Pontifical Academy for Life, 2020, 7). For one, education must be made available to the elderly, persons with disabilities, and other marginalized groups, which would only make them more independent. So, in the context of AI systems, this means that these technologies must be used for inclusive education, providing opportunities for everyone to participate in societal activities.

The area on education also stresses the value of AI-enriched curricula, so that the basic principle that no one gets left behind comes into fruition (Pontifical Academy for Life, 2020, 8). These school programs must ensure that the younger generations would be able to "fully express their capabilities and [work] for the good of the community, even when there is no personal benefit to be gained" (8). By developing these educational programs, AI-propelled opportunities related to social inclusion, individual respect, and the like would be further magnified, while also highlighting the potential critical issues that intelligent technologies might produce.

5.2.3 Rights

The third impact area, rights, encourages stakeholders to consciously reflect on, if not codify in pertinent legal regulations, values, and principles, the view that AI technologies should be made and employed for humanity, if not the entire world (Pontifical Academy for Life, 2020, 9). It further emphasizes the idea that any AI initiatives must always come with robust digital security measures. This is because these types of measures would ensure that AIs would be for the benefit of everyone, acting as conduits of international peace.

Furthermore, also stated in the impact area on rights is the notion that

> [i]n order for AI to act as a tool for the good of humanity and the planet, we must put the topic of protecting human rights in the digital era at the heart of public debate. The time has come to question whether new forms of automation and algorithmic activity necessitate the development of stronger responsibilities. In particular, it will be essential to consider some form of "duty of explanation": we must think about making not only the decision-making criteria of AI-based algorithmic agents understandable, but also their purpose and objectives. (Pontifical Academy for Life, 2020, 9).

Finally, the impact area on rights underscores the importance of transparency, traceability, and responsibility in computer-aided decision-making processes, wherein people must be able to understand the logic behind AI algorithms. This echoes some of the concerns related to what has come to be known as the "AI black box problem", which Xu (2019) elucidates as follows:

> This black-box phenomenon causes users to question the decisions from the system.... This reflexive skepticism directly affects users' trust and decision-making efficiency, thus also affecting the adoption of AI solutions. Because of the black-box effect, AI solutions are not explainable and comprehensible to users. This phenomenon may occur across various types of AI applications, including financial and legal decisions, medical diagnoses, industrial process monitoring, security screening, employment recruitment, legal judgment, university admissions, smart homes, and autonomous vehicles. (Xu, 2019, 43–44)

In light of this, better forms of regulations, like those that clearly nurture transparency and other ethical principles, must be employed to more complex AI systems, especially the ones that might ensue human rights issues (Pontifical Academy for Life, 2020, 9).

In the end, the "Rome Call for AI Ethics" maintains that, in order to properly address the impact areas mentioned above, all concerned stakeholders (i.e., international organizations, non-governmental agencies, political leaders, researchers, practitioners, etc.) should embrace an "algor-ethical" vision—"an approach of ethics by design" (Pontifical Academy for Life, 2020, 10). The same document further claims that the said approach may be achieved via considering the following principles:

- Transparency: in principle, AI systems must be explainable
- Inclusion: the needs of all human beings must be taken into consideration so that everyone can benefit and all individuals can be offered the best possible conditions to express themselves and develop
- Responsibility: those who design and deploy the use of AI must proceed with responsibility and transparency
- Impartiality: do not create or act according to bias, thus safeguarding fairness and human dignity
- Reliability: AI systems must be able to work reliably
- Security and privacy: AI systems must work securely and respect the privacy of users. (Pontifical Academy for Life, 2020, 10)

5.3 Artificial Intelligence and the Philippines

In general, the Philippines has had a handful of AI-related efforts and initiatives coming from the national government, private sector, and the academe, to name a few. For instance, the national government, via the Department of Trade and Industry (DTI), rolled out its AI Roadmap in May 2021 (Campos, 2021). Supposedly, this roadmap hopes to position the Philippines as an "AI Center for Excellence" in the region. In order to attain this vision, the same document also specifies key strategies.

The AI Roadmap of DTI cites several pillars and dimensions (Department of Trade and Industry, 2021, 8). Supposedly, these pillars and dimensions would aid and guide the Philippines in terms of having clearly defined metrics in assessing its AI implementation progress. Hence, if the goal of the country is to become one of the competitive AI-driven economies, the rubrics would help ensure that it is on the right track.

In terms of the pillars, they are further classified into two kinds, namely: (1) implementation and (2) innovation (Department of Trade and Industry, 2021, 8). On the one hand, the implementation pillar looks into how AI technologies are operationalized by practitioners in the areas of business, government, and the community. The innovation pillar, on the other hand, essentially concerns itself with "technology breakthroughs and advancements in methodology that are indicative of greater capacity for AI in the future" (8–9).

As regards the dimensions that would guide the Philippines in evaluating its AI implementation progress, DTI's AI Roadmap states the following:

> The *first* dimension is **Digitization and Infrastructure** as it is critical in achieving all the strategic goals defined in this Roadmap. The *second* dimension is **Research and**

> **Development** that includes as subset "Investing in Strategic Sectors" and "International Collaboration." The *third* dimension is the **Workforce**, which focus[es] on anticipating and preparing for the potential impact of AI technologies by ensuring enough talents and skilled practitioners that can deliver AI solutions (more on application and implementation) and perform AI research (academic and industry-driven). This dimension will also cover both reskilling and upskilling of the workforce. Finally, the *fourth* dimension is **Regulation**, which protects human rights and puts into place equitable AI activating opportunities. (Department of Trade and Industry, 2021, 10)

In addition, the same document qualifies that Digitalization and Infrastructure, Workforce Development, and Regulation dimensions may be classified under the implementation pillar. Meanwhile, the Research and Development dimension falls under the pillar of innovation.

By considering the pillars and dimensions identified in the AI Roadmap of DTI, along with certain strategic imperatives and tasks (Department of Trade and Industry 2021, 9–14), the Philippines would be on pace to attain its goal of being one of the more competitive AI-driven economies worldwide (8). Former DTI secretary Ramon M. Lopez also explains that "through the AI Roadmap, we hope to establish the Philippines as an 'AI Center for Excellence' in the region that is backed by our rich local talent pool and vibrant innovation and entrepreneurship ecosystem. Our country can also serve as a big data processing hub providing high-value data analytics and AI services to the world, and this can become our next higher value, higher paying service-winner after the BPO industry" (Campos, 2021).

Furthermore, the Philippine national government has also embarked on the project of putting up the Center for AI Research (CAIR)—the country's first center for researchers, data scientists, and the like (Cos, 2022). This center aims to foster collaborations among these practitioners on AI-related research and development services, including consulting. Moreover, this hub is also intended to be a place for the said professionals to develop AI products and facilitate data literacy programs. For its part, DTI envisions the Center for AI Research to focus on AI projects related to precision farming, smart manufacturing, and business process outsourcing, among others.

As for the AI-related efforts from the private sector, several companies have also closely worked with the Philippine government, like in the establishment of CAIR. For one, Crismundo (2022) details that a number of businesses in the Philippines have already pledged to invest around USD20 million to put up the said center. Among the companies that have entered into an agreement with DTI for the CAIR project include Global Learning Solutions and UnionBank of the Philippines (Ochave, 2022).

Furthermore, Cahiles-Magkilat (2022) reports that more than 50 technology startups would also aid the Philippines in forging its AI ecosystem, projected to contribute about USD92 billion by 2030 once this ecosystem is fully developed. The same report further contextualizes that "[o]ne of the DTI's flagship initiatives that would drive digital transformation is the establishment of an Industry 4.0 Pilot Factory, which will host pilot, demonstration, and learning laboratories for Industry 4.0 technologies, such as, robotics, intelligent manufacturing systems, and cyber-physical systems. It will serve as a technology platform for various stakeholders and a training and research hub where industries can have hands-on experience on Industry 4.0 applications" (Cahiles-Magkilat, 2022).

In terms of the initiatives from the academe (i.e., in relation to the development and adoption of AIs in the Philippines), Concepcion et al. (2019) already came up with key AI policy recommendations that highlight the following four sectors, namely: (1) labor and employment, (2) human resource development and education, (3) government policies and laws, and (4) research and development and innovation. Supposedly, by focusing on addressing the AI-related issues in these areas, the Philippines would be positioned to maximize the positive effects of these advanced technologies.

With regard to labor and employment, Concepcion et al. (2019, 7) hold that AIs should be treated by policy makers, business executives, and the like as actual platforms for economic development. Additionally, they maintain that, in terms of human resource development and education, the primary focus of the Philippines should be the creation of AI-based educational programs. Among the programs include technology-based ones that would hone the digital skills of students.

As for the government policies and laws sector, Concepcion et al. (2019, 7) recommend that the Philippines also forges partnerships with global AI corporations, like what other nations (e.g., India, United Arab Emirates, etc.) have already done. Finally, they also advise that the Philippine government develops an AI ecosystem that is robust for big data. Note that, apart from the AI policy suggestions proffered by Concepcion et al. (2019), others have also proposed their own set of recommendations.

In order to harness and further develop the applications of AI systems in the economy, industry, society, and government, Galvez et al. (2019) also offer their own set of AI policy suggestions. In a way, it may be said that their recommendations are arguably reminiscent of those put forward by Concepcion et al. (2019). For one, the latter explain that AI technologies have

> lots of applications in the field of machine learning, natural language processing, robotics, and expert systems. It offers positive impacts on the economy, society, industry, and government. To maximize this, the government must create a national policy that will focus on the security of AI systems and support for research and development. Informing the public about… AI will also help them to understand the technology and how to use it safely. (Galvez et al., 2019, 4)

In the same vein, Rosales et al. (2020) propose the same kind of AI policy suggestions, which would aid the Philippines in fully harnessing and capitalizing on the benefits of computational technologies.

In light of the AI-related efforts and initiatives from the national government, private sector, and the academe mentioned above, it could be said that, in general, the Philippines is not that "in the dark" in terms of adapting to the challenges brought forth by advanced intelligent systems. However, with regard to responding to the "Rome Call for AI Ethics", one may argue that answering this call from the Vatican is not that straightforward of an endeavor. For one, as mentioned earlier, the Philippines has had problems in responding to pronouncements issued by the global community, like those related to climate change (La Viña & Reyes, 2021), human rights issues (Pernia, 2019; Regilme, 2021), and the like.

Following the strategy of using a set of conditions as preliminary assessment indicators to check a state's capacity of addressing large-scale problems (Boyles et al., 2022),[1] this paper argues that the Philippine government is not that primed to answer the Vatican's call (i.e., in light of its track record of handling concerns related to ethics, education, and rights). The study makes use of the three impact areas included in the "Rome Call for AI Ethics", employing them as initial measures, and examines the present status of such domains in relation to the Philippine context. The conjecture is that, by considering key government undertakings in the past, the level of importance that the Philippines has generally accorded the areas of concerns would be brought to light. Thus, if certain government acts or initiatives seem to exemplify that they run counter to the ideals espoused in the said call, then it may be said that formulating any concrete response would pose a challenge.

5.4 Analyzing the Impact Areas in the Philippines

With regard to the first area identified by the "Rome Call for AI Ethics", ethics, recall that this focuses on the idea that AI systems should be conceptualized, modeled, and mainly used to cater to the needs and interests of human beings, including their living environments (Pontifical Academy for Life, 2020, 5). This is because all individuals are assumed to be free and equal, specifically in terms of rights and dignity. So, the rights and freedoms of everyone should always be safeguarded whenever one develops and employs AIs, while also giving much importance to the concerns of those most vulnerable (Pontifical Academy for Life, 2020, 6). In light of the characterization of this impact area, it may be said that such accentuates the view that the fundamental human rights of individuals, as well as the upkeep of the environment, ought to be upheld at all times.

In philosophical terms, the concept of "human rights" may be understood as the moral rights of human beings (Evangelista & Mabaquiao, 2020, 23). Basically, this means that the possession of this kind of rights largely depends on one being considered a moral person—someone who has moral status or standing, making them appropriate objects of ethical concern (Evangelista & Mabaquiao, 2020, 21). It must also be kept in mind that "one becomes a moral person, and thus acquires moral right when one possesses the defining qualities of moral personhood, which may include sentience (the capacity to experience pleasure and pain) and rationality (the

[1] Citing research scientist Victoria Krakovna's (2020) suggestion that the three problem areas related to (1) learning from experience, (2) warning signs, and (3) consensus on the problem may be considered in preparing for a slow AI takeoff, especially that these areas have become useful in assessing the COVID-19 pandemic response of the global community, Boyles et al. (2022) argue that the Philippine disaster response towards a technological singularity needs to be worked on.

capacity to know and choose freely)" (Evangelista & Mabaquiao, 2020, 23).[2] Unfortunately, if one considers the previous undertakings or efforts made by the Philippine national government to safeguard human rights, things become quite concerning.

In terms of handling human rights issues, the reputation of the Philippines has been considered infamous the last few years. For instance, non-profit organization Freedom House has scored the country as only "partly free" (scoring 55 out of 100) in 2022, specifically scoring 25 out of 40 for political rights, and 30 out of 60 for civil liberties (Freedom House, 2022). Among the developments factored in from the previous year for the score include the supposed hostile stand taken by the administration of former Pres. Rodrigo R. Duterte towards higher education institutions (HEIs) and academics that teach materials with subversive ideologies, the pronouncement of the same head of state that all communists in the country must be eradicated by the military and police, which has allegedly led to the death of nine activists and arrest of six others in provinces outside the nation's capital in 2021, and so on.

Furthermore, with regard to observing due process in civil and criminal matters, Freedom House (2022) scored the Philippines 0 out of 4 in their 2022 report. This is because of the following:

> The justice system fails to guarantee due process rights. Arbitrary detention, disappearances, kidnappings, and abuse of suspects are common. The [Office of the United Nations High Commissioner for Human Rights] catalogued numerous due process violations linked to Duterte's war on drugs, including the use of watch lists compiled by local officials that identify targets for home "visitations" that do not require warrants and often lead to extrajudicial execution.
>
> The police and military have been implicated in corruption, extortion, and involvement in the illegal drug trade.
>
> The Anti-Terrorism Act that took effect in July 2020 provides the state significant powers, including the warrantless arrest and detention of people designated as terrorists by an Anti-Terrorism Council appointed by the president. Rights advocates sharply criticized the law's broad definition of terrorism as endangering dissent and free speech and stated that limits incorporated into the law were too vague to offer effective protection. At least 37 petitions challenging the law's constitutionality have been filed in the Supreme Court. In December 2021, the Supreme Court upheld the constitutionality of the Anti-Terrorism Act. (Freedom House, 2022)

So, it may be inferred that, in terms of upholding the fundamental human rights of its people, the Philippine national government is far from reaching this ideal. The same sentiment may also be derived in other studies, like in Pernia's (2019) depiction of the Duterte administration.

---

[2] Though there are still ongoing discussions on which theory of moral personhood is actually correct (Evangelista & Mabaquiao, 2020, 26–30), this concern is well beyond the scope of the present study.

Similarly, the apparent characterization of non-governmental organization Amnesty International of the Philippines is not that appealing in terms of safeguarding the human rights of people. For one, the group cites the lack of accountability of the country's national government, specifically in condoning several human rights violations, including unlawful killings (Amnesty International, 2022). Allegedly, this is most glaring in former Pres. Duterte's "war on drugs" campaign, which has prompted the International Criminal Court's (ICC) pre-trial chamber to conduct a probe. The same administration, however, declined to cooperate with the ICC investigation.

Amnesty International (2022) also claims that the rights of indigenous people in the Philippines, as well as activists fighting for them, have been trampled on by the government. For instance, the group reports that, sometime in February 2021, "security forces raided a school in Cebu City in which students from Indigenous communities displaced by armed conflict were living. Seven people, including students, teachers and a community elder, were arrested without warrants and charged with kidnapping and human trafficking" (Amnesty International, 2022). The people were purportedly detained for around 3 months before the formal charges were officially dismissed.

Thus, in terms of guarding the rights of the most vulnerable people, it seems that the Philippine national government is quite underperforming. Other instances that could somehow provide a snapshot on how the country fairs against the ideals espoused in the impact area on ethics include the Philippines placing 158th place out of 180 countries in the 2022 Environmental Performance Index—a list that ranks "countries on their progress toward improving environmental health, protecting ecosystem vitality, and mitigating climate change" ("Philippines places 158th out of 180 in environmental sustainability ranking", 2022). Also, following the declaration of the United Nations that Internet access must be treated a basic human right (Estes, 2011), it may be said that the Philippine government has not given much importance to such, since the country has one of the slowest Internet connection speeds (Salac & Kim, 2016, 80–81).

If one grants that the main focus of the area on ethics is upholding the basic human rights of individuals, including the proper care of the planet (Pontifical Academy for Life, 2020, 5), it seems that the Philippines has its back against the wall. This is because of how its national government has handled issues related to these kinds of concerns. Thus, addressing this specific domain in the "Rome Call for AI Ethics" would not be that easy for the country.

As regards the impact area on education, recall that the primary focus of such is the creation of curricula or school programs that highlight social inclusion- and individual respect-related opportunities powered by AI systems (Pontifical Academy for Life, 2020, 8). At the same time, this area also emphasizes the importance of discussions regarding the potential issues that advanced intelligent technologies might pose. This impact area also advances the democratization of education for all human beings, wherein no one should be left behind such as the elderly, persons with disabilities, and other marginalized groups.

So, it may be said that, in the "Rome Call for AI Ethics", the impact area on education underlines the idea that AI technologies must be used for the attainment of inclusive education; these technologies must provide equal opportunity to everyone. Note that this also relates to the idea that individuals have the right to education, regardless if the said right is considered a positive or negative one (Evangelista & Mabaquiao, 2020, 22).[3] Regrettably, it appears that the Philippine government has constantly struggled to cope up with the challenges related to the domain of education.

In depicting the status of the education sector in the Philippines, Durban and Catalan (2012) hold that:

> Politics in education is an issue that presently pervades [the] educational system in the country. The government, specifically the legislators, is inept in formulating laws that can address the crisis in the educational system. A sad reality that is happening right now is the formulation of policies with the main purpose of making our educational system at par with those in other countries, but there are no concrete guidelines as to how these are to be implemented. Most educational experts are technocrats with no experience in the field. Yes, their programs are good, to say the least, but because of their lack of experience in actual classroom teaching, they fail to study the application of these programs. (Durban & Catalan, 2012, 67–68)

This characterization may be further seen if one cites a handful of instances wherein the Philippines tried to implement educational reforms, if not adapt to the call of the times.

Prior to adopting the K-12 program, which "covers kindergarten, six years of primary education, four years of junior high school, and two years of senior high school" (Parrocha, 2023), the Philippines was one of the last few nations to embrace this type of basic education system. So, before implementing the K-12 scheme in 2012, the Philippines, along with Angola and Djibouti, were the only ones that had a basic education system that lasted for 10 years (Pazzibugan, 2013); the Philippines lagged behind the rest of the world. Sadly, after over 10 years of implementation, Vice Pres. Sara Duterte, who concurrently acted as Department of Education (DepEd) secretary, maintains that the education system has had failures and has even become a burden for teachers (Bautista & Aning, 2023).

In her State of Education Report in January 2023, Vice Pres. Duterte said that, after assessing the country's K-12 curriculum, there is "insufficient knowledge [among teachers] on developing twenty-first-century skills, including higher-order thinking skills among learners" (Bautista & Aning, 2023). She further explains that these failures cannot really be attributed to the basic education teachers, since it is the "system that burdens them with backbreaking and time-consuming administrative tasks, a system that provides no adequate support and robs them of the

[3] Evangelista and Mabaquiao (2020, 22) clarify that "[t]he right to education, for instance, in consideration of the duty that it imposes on the state, is negative or positive depending on the economic condition of the state. If the state, given its economic condition, can supply the resources needed for its citizens to avail of this right, then it is just proper that such a right be treated by the state as a positive one".

opportunity to professionally grow and professionally teach, assist, and guide our learners."

Vice Pres. Duterte also elucidates that the lack of proper school facilities and infrastructures still remains one of the main problem areas in the Philippines (Ombay, 2023). For one, she mentions that only 104,536 school buildings are in good working condition (i.e., out of the 327,851 total across the country). Vice Pres. Duterte also claims that "100,072 school buildings… need minor repairs, 89,252… need major repairs, and 21,727… are set for condemnation" (Ombay, 2023).

In addition, Vice Pres. Duterte reports that most school infrastructures in the Philippines are not really built to withstand natural calamities (Ombay, 2023). The implication of this is that the facilities have always been prone to disasters, like earthquakes, typhoons, and so on. As early as 2005, the World Bank issued a report entitled *Natural Disaster Risk Management in the Philippines Reducing Vulnerability*, which shows how vulnerable the country is to these kinds of disasters. This report states that the Philippines is "one of the most natural hazard prone countries in the world" (World Bank, 2005, S-1). DepEd could have seriously considered this almost archaic, yet crucial, piece of information in disaster-proofing the country's school infrastructures and facilities way before.

Hence, in light of the instances mentioned above, it could be said that the Philippines has always struggled in implementing important educational reforms. Actually, this idea may also be said about how the country responded to the education-related challenges stemming from the COVID-19 pandemic. For one, the Philippines and Venezuela were the last two countries that allowed the face-to-face resumption of classes in schools (Deiparine, 2021).

If the central concern of the impact area on education is the development of school programs or curricula that highlight social inclusion- and individual respect-related opportunities aided by AI systems (Pontifical Academy for Life, 2020, 8), then it seems that the Philippine national government would be in a tough position in realizing this ideal. Even the thought of democratizing the access to education in the Philippines appears to be difficult to actualize in light of its undertakings discussed above. Thus, with regard to the impact area on education in the "Rome Call for AI Ethics", it seems that the Philippines would be quite challenged in appropriately responding to such.

Recall that the third impact area highlighted in the "Rome Call for AI Ethics" is about rights (Pontifical Academy for Life, 2020, 9). In general, this area encourages stakeholders to reflect on, if not codify in pertinent legal regulations, principles, and values, the idea that AIs must be made and used for humanity, if not the entire world. This means that any AI-related initiatives should always come with robust digital security measures to ensure that these systems would be for the benefit of humanity and the planet. So, the third impact area on rights underlines the value of transparency, traceability, and responsibility in computer-aided decision-making processes; people should always be able to understand the logic behind AI algorithms.

Philosophically speaking, issues concerning transparency, traceability, and responsibility could be related to the concept of justice, specifically the procedural type. Procedural justice "refers to the fairness in the process by which a decision is made, especially in resolving disputes and allocating resources. It is usually associated with the legal concept of due process" (Evangelista & Mabaquiao, 2020, 123). Moreover, in terms of the responsible use and development of AIs, issues regarding such have also been brought up in connection with the notion of moral responsibility. Noorman (2023) explains that "[a]s computer technologies have become a more integral part of daily activities, automate more decision-making processes and continue to transform the way people communicate and relate to each other, they have further complicated the already problematic tasks of attributing moral responsibility".

Accordingly, transparency, traceability, and responsibility are important values that one needs to take into account in light of AIs. It is quite unfortunate, however, that if one considers how the Philippine government has handled issues related to transparency, accountability, and the like in the past, there are reasons to believe that it would not be able to fully safeguard these things. This means that the country would not really be able to meet the ideals espoused by the impact area on rights.

Citing Varela (2003), Torneo and Mojica (2020) contend that transparency in the Philippines has always been impeded by graft and corruption. They also maintain that government laws and regulations designed to mitigate the latter are not that effective, if not they conflict with one another. The government transparency- and corruption-related issues in the Philippines have been quite common through the years, repeatedly haunting the country and its citizens from one administration to the next. Consider, for instance, the anti-corruption efforts by the last three administrations before the incumbent, Pres. Ferdinand Marcos, Jr.

In characterizing the administration of former Pres. Gloria Macapagal-Arroyo (i.e., in its efforts to fight corruption), Batalla (2015, 51) maintains that it had a "curious relationship with corruption". This is because the Arroyo administration, on the one hand, tried to institute several anti-corruption programs, while also initiating key reforms such as amending procurement laws, indicting corrupt government and military officials, and so on (Oyamada, 2005). On the other hand, however, the same administration also faced a number of corruption scandals, like those related to the 2004 Philippine elections, Fertilizer Fund scam, and NBN-ZTE broadband controversy, among others (Batalla, 2015, 51–52). So, it may be contended that in spite of the seemingly pioneering initiatives of the Arroyo regime to combat corruption, the latter still permeated its different government agencies.

Furthermore, after analyzing the anti-corruption initiatives of the administration of former Pres. Benigno Aquino III, Batalla (2015) concludes that:

> Legal deficiencies and law enforcement difficulties plague the anti-corruption crusade in the Philippines. Unresponsive laws and poor enforcement reflect the deficient character of state institutions, whose continued weakness and malleability provide tremendous opportunities for the private accumulation of wealth and power, especially for those with favorable access to state resources. The preservation of weak state institutions follows the logic of patrimonial domination in the Philippines. *Ceteris paribus*, one can expect very limited success in the overall fight against corruption. (Batalla, 2015, 70)

So, even before the Aquino government ended, there were already some doubts that its anti-corruption campaigns would really succeed. Those government agencies that have been identified as prone to corruption are, in fact, the ones that possess "stiff and protracted resistance against reform initiatives" (Batalla, 2015, 71).

Finally, as regards the anti-corruption campaigns of the Duterte administration, Kossow (2019, 8) explains that "Duterte's advocacy against corruption was not matched by his actions as president". This is because Pres. Duterte's anti-corruption assertions seem to run counter against his actions. For one, Kossow (2019) cites Pres. Duterte's alleged act of not including certain bank accounts as part of his asset declarations while running for the presidency (Ong, 2016). Citing Calonzo (2018), Kossow (2019, 5) additionally highlights the occasion wherein Pres. Duterte re-hired a number of government officials whom he had previously fired due to corruption allegations. So, for Kossow (2019), former Pres. Duterte's actions arguably exemplify his failed attempts to bring corruption to an end.

Considering how the administrations cited above tried to handle, if not fight, corruption-related concerns in the past, it may be said that things are not that straightforward when it comes to curbing this problem. So, in relation to the third impact area in the "Rome Call for AI Ethics" that focuses on the idea of protecting the rights of everyone in light of AI technologies, it appears that the Philippine government would be quite challenged in satisfying this domain. If the government has perennially failed to put much importance in transparency, responsibility, and the like in the past (i.e., without even taking into account the disruptive effects of AIs), then it would likely find itself in a tough situation once these advanced technologies are factored in.

Though preliminary and non-exhaustive, the instances specified earlier as regards the particular undertakings, actions, or initiatives of the Philippine government somehow show the level of importance that it has generally accorded the areas on ethics, education, and rights, which have been underscored in the "Rome Call for AI Ethics". In a way, these facts arguably provide an initial, yet practicable, information that any prospective response to the said call should account for. If one considers the track record of the Philippine government in handling issues related to ethics, education, and rights in the past, then it may be contended that a lot of work has to be done first before it successfully, or perhaps even just decently, answers the Vatican's call. Nevertheless, it seems that, though there might be a number of challenges in reforming the standing of these impact areas in the Philippines, there are also indications that things are moving towards a more positive direction.

For one, as regards the area on ethics, the pronouncement provided by the Office of the Solicitor General (under the current government of Pres. Ferdinand Marcos, Jr.) that it plans to file an appeal to the International Criminal Court regarding the reopening of the latter's drug war probe on the Duterte administration has been seen by a number of observers as a good sign (Gavilan, 2023; Navallo, 2023). This is because, as compared to Pres. Duterte's stance that the government would never cooperate with any foreign-led investigation, the incumbent regime's participation in standard ICC processes may be deemed a relatively welcome development in itself.

Moreover, in terms of education, the Philippine national government has also taken a few steps to incorporate AIs into certain programs. For example, DepEd has partnered with private companies, like Microsoft, to foster "inclusive and accessible education for learners with disabilities (LWDs) through the use of technology" (Sevillano, 2022). Additionally, it also helps that the local church, through the Catholic Bishops' Conference of the Philippines (CBCP), has acknowledged the value of the relationship between education and AIs. Responding to the Senate of the Philippine's proposal to mandate all elementary and high schools (i.e., both public and private) to include a Good Manners and Right Conduct (GMRC) course into school curricula, CBCP Episcopal Commission on Youth chairperson Bishop Rex Andrew Alarcon said that:

> I welcome this move. GMRC is actually at the core of education. At the time of digital transformation and AI (Artificial Intelligence), all the more we realize the need to focus on being more human. Moreover, at the heart of legislation and its implementation is the link between laws and values.... If the citizens and the community do not see the values behind the laws, laws become useless. Institutionalizing GMRC in education hopefully will reinforce the message that real progress can only be achieved by forming the human person first and foremost.... It does not stop in schools, however, it is a lifelong task. It is learning for life (Patinio, 2020).

Lastly, with respect to the impact area on rights, Baclig (2023) reports that, although the country still maintained a low score, the Philippines modestly improved in ranking in Transparency International's 2022 Corruption Perception Index. In the 2022 version of the index, the Philippines placed 116th out of 180 countries—a slight improvement from its 117th rank the previous year. Though some might argue that the said advancement is quite marginal, it is still some form of furtherance nonetheless.

So, if one considers the said moderate improvements in the Philippines (i.e., with respect to the impact areas identified in the "Rome Call for AI Ethics"), it may be argued that there are some indications that things are moving towards a more positive direction. The developments might not be night and day, but at least there is a bit of hope to hold onto. Hope has always been a good thing.

5.5 Conclusion

In this article, it has been demonstrated that, in light of how it has generally addressed issues related to the impact areas indicated in the "Rome Call for AI Ethics", the Philippine national government would be challenged in responding to the Vatican's call. By citing the areas related to ethics, education, and rights, using them as preliminary assessment measures to check the state's capacity to respond to the said call, it may be said that the Philippine government is not ready to answer such given its past performances in handling issues related to these domains. Nevertheless, it was also shown that, though there might be several challenges in reforming the standing of the three impact areas in the Philippines, there are some indications that things are moving towards a positive direction.

Considering that the cited anecdotes, including the general analysis, in the present study are still quite preliminary and non-exhaustive, more research needs to be done to fully assess the preparedness of the Philippines in answering the "Rome Call for AI Ethics". For example, some, like Haeberer et al. (2021), have suggested examining the diverse assessment tools used in measuring the preparedness of a nation in facing a public health emergency. So, perhaps studies that analyze the different assessment measures on the AI readiness of countries could be undertaken, if not research that center on the development of new evaluation tools in themselves.

Furthermore, there are those such as Selvamani (2022) that have also claimed that the Catholic church needs to do more beyond the issuance of the "Rome Call for AI Ethics". Selvamani (2022, 58) contends that, "[t]hough the Vatican is making effort to engage technological progress through these events, these will remain only as a photo-op moment, unless the faithful and the educational, social and health-care institutions to actively lead the dialogue between the stakeholders". Hence, it would also be interesting to critically examine the actual role that the church would have to play in the preparations for the coming of highly advanced intelligent systems.

References

Amnesty International. (2022). Everything you need to know about human rights in Philippines. *Amnesty International Report 2021/22*. https://www.amnesty.org/en/location/asia-and-the-pacific/south-east-asia-and-the-pacific/Philippines/report-Philippines/. Accessed 14 March 2024.

Baclig, C. E. (2023, February 1). PH inches up in corruption index but low score persists. *Inquirer.net*. https://newsinfo.inquirer.net/1724067/for-posting-edited-ph-inches-up-in-corruption-index-but-low-score-persists. Accessed 13 March 2024.

Batalla, E. V. C. (2015). Treading the straight and righteous path: Curbing corruption in The Philippines. *Asian Education and Development Studies, 4*(1), 51–75. https://doi.org/10.1108/AEDS-10-2014-0043

Bautista, J., & Aning, J. (2023, January 31). PH education woes not teachers' fault, says VP Duterte. *Inquirer.net*. https://newsinfo.inquirer.net/1723204/ph-education-woes-not-teachers-fault-says-vp. Accessed 14 March 2024.

Boyles, R. J. M., Dacela, M. A., Evangelista, T. R., & Rodriguez, J. C. (2022). COVID-19 and singularity: Can The Philippines survive another existential threat? *Asia-Pacific Social Science Review, 22*(2), 181–195.

Cahiles-Magkilat, B. (2022, September 11). PH now hosts over 50 tech startups — DTI. *Manila Bulletin*. https://mb.com.ph/2022/09/11/ph-now-hosts-over-50-tech-startups-dti/. Accessed 14 March 2024.

Calonzo, A. (2018, July 4). Duterte's anti-corruption purge now revolving door of officials. *Bloomberg*. https://www.bloomberg.com/news/articles/2018-07-03/duterte-s-anti-corruption-purge-now-revolving-door-of-officials#xj4y7vzkg. Accessed 13 March 2024.

Campos, O. V. (2021, May 5). DTI launches AI roadmap. *Manila Standard*. https://manilastandard.net/business/economy-trade/353646/dti-launches-ai-roadmap.html. Accessed 12 March 2024.

Concepcion, R. S., Bedruz, R. A. R., Culaba, A. B., Dadios, E. P., & Pascua, A. R. A. R. (2019). The technology adoption and governance of artificial intelligence in The Philippines. In *2019 IEEE 11th International Conference on Humanoid, Nanotechnology, Information Technology, Communication and Control, Environment, and Management (HNICEM)* (pp. 1–10). IEEE. https://doi.org/10.1109/HNICEM48295.2019.9072725

Cos, W. (2022, August 22). PH AI research center to open in September. *ABS-CBN News*. https://news.abs-cbn.com/business/08/22/22/ph-ai-research-center-to-open-in-september. Accessed 12 March 2024.

Crismundo, K. (2022, June 26). Conglomerates investing $20-M for AI center. *Republic of the Philippines: Philippine News Agency*. https://www.pna.gov.ph/articles/1177557. Accessed 14 March 2024.

Deiparine, C. (2021, September 8). Only Philippines, Venezuela yet to allow return to classroom learning. *Philstar.com*. https://www.philstar.com/headlines/2021/09/08/2125759/only-philippines-venezuela-yet-allow-return-classroom-learning. Accessed 13 March 2024.

Department of Trade and Industry. (2021). National artificial intelligence strategy roadmap for The Philippines. *Republic of the Philippines: Department of Trade and Industry*. https://drive.google.com/file/d/1de5kfaGi3tdUgxu1UPV8iRxKpXECTkDp/view. Accessed 14 March 2024.

Durban, J. M., & Catalan, R. D. (2012). Issues and concerns of Philippine education through the years. *Asian Journal of Social Sciences & Humanities, 1*(2), 61–69.

Estes, A. C. (2011, June 6). The U.N. declares internet access a human right. *The Atlantic*. https://www.theatlantic.com/technology/archive/2011/06/united-nations-wikileaks-internet-human-rights/351462/. Accessed 14 March 2024.

Evangelista, F. J. N., & Mabaquiao, N. M., Jr. (2020). *Ethics: Theories and applications*. Anvil Publishing, Inc..

Freedom House. (2022). Philippines: Freedom in the World 2022 Country Report. *Freedom House*. https://freedomhouse.org/country/philippines/freedom-world/2022. Accessed 12 March 2024.

Galvez, R. L., Culaba, A. B., Dadios, E. P., & Bandala, A. A. (2019). Artificial intelligence: Policy paper. *Journal of Computational Innovations and Engineering Applications, 4*(1), 1–5.

Gavilan, J. (2023, January 27). Rights groups see new light towards justice in ICC drug war probe resumption. *Rappler*. https://www.rappler.com/nation/human-rights-groups-statements-icc-drug-war-probe-resumption/. Accessed 14 March 2024.

Haeberer, M., Tsolova, S., Riley, P., Cano-Portero, R., Rexroth, U., Ciotti, M., & Fraser, G. (2021). Tools for assessment of country preparedness for public health emergencies: A critical review. *Disaster Medicine and Public Health Preparedness, 15*(4), 431–441. https://doi.org/10.1017/dmp.2020.13

Heinrichs, B. (2022). Discrimination in the age of artificial intelligence. *AI & SOCIETY, 37*, 143–154. https://doi.org/10.1007/s00146-021-01192-2

Hyken, S. (2017, June 10). Half of people who encounter artificial intelligence don't even realize it. *Forbes*. https://www.forbes.com/sites/shephyken/2017/06/10/half-of-people-who-encounter-artificial-intelligence-dont-even-realize-it/?sh=6c32d421745f. Accessed 14 March 2024.

Kossow, N. (2019). *Populism and corruption*. Transparency International Anti-Corruption Helpdesk Answer. Transparency International. https://knowledgehub.transparency.org/assets/uploads/helpdesk/populism-and-corruption-2019-final.pdf

Krakovna, V. (2020, May 31). Possible takeaways from the Coronavirus pandemic for slow AI takeoff. *Victoria Krakovna*. https://vkrakovna.wordpress.com/2020/05/31/possible-takeaways-from-the-coronavirus-pandemic-for-slow-ai-takeoff/. Accessed 13 March 2024.

La Viña, A. G. M., & Reyes, J. J. M. (2021). Innovation and dysfunction: Three decades of climate change governance in The Philippines. In J. Marquardt, L. L. Delina, & M. Smits (Eds.), *Governing climate change in Southeast Asia: Critical perspectives* (pp. 133–147). Routledge. https://doi.org/10.4324/9780429324680

Lubov, D. C. (2023, January 10). Pope: AI Ethics must safeguard the good of human family. *Vatican News*. https://www.vaticannews.va/en/pope/news/2023-01/pope-francis-receives-rome-call-vatican-audience.html. Accessed 14 March 2024.

Navallo, M. (2023, February 10). PH appealing ICC drug war probe 'good sign,' says rights researcher. *ABS-CBN News*. https://news.abs-cbn.com/video/news/02/10/23/ph-appealing-icc-drug-war-probe-a-good-sign-hrw. Accessed 12 March 2024.

Noorman, M. (2023). Computing and moral responsibility. In E. N. Zalta (Ed.), *The Stanford encyclopedia of philosophy*. https://plato.stanford.edu/entries/computing-responsibility/. Accessed 12 March 2024

Ochave, R. M. D. (2022, September 12). UnionBank, GLS sign on as gov't partners in AI research center. *BusinessWorld Online*. https://www.bworldonline.com/economy/2022/09/12/474026/unionbank-gls-sign-on-as-govt-partners-in-ai-research-center/. Accessed 14 March 2024.

Ombay, G. (2023, January 30). VP Sara: Lack of school facilities still basic education's main problem. *GMA News Online*. https://www.gmanetwork.com/news/topstories/nation/859083/vp-sara-lack-of-school-facilities-still-basic-education-s-main-problem/story/. Accessed 14 March 2024.

Ong, W. (2016, April 26). Duterte SALNs don't match alleged bank accounts. *Rappler*. https://www.rappler.com/newsbreak/in-depth/131167-rodrigo-duterte-bank-accounts-saln/. Accessed 13 March 2024.

Oyamada, E. (2005). President Gloria Macapagal-Arroyo's anti-corruption strategy in The Philippines: An evaluation. *Asian Journal of Political Science, 13*(1), 81–107. https://doi.org/10.1080/02185370508434251

Parrocha, A. (2023, January 10). DepEd to give updates on K-12 curriculum on Jan 30. *Republic of the Philippines: Philippine News Agency*. https://www.pna.gov.ph/articles/1192378. Accessed 14 March 2024.

Patinio, F. (2020, February 18). CBCP backs Senate nod on GMRC bill. *Republic of the Philippines: Philippine News Agency*. https://www.pna.gov.ph/articles/1094143. Accessed 13 March 2024.

Pazzibugan, D. Z. (2013, June 3). 'K + 12' still struggling. *Inquirer.net*. https://newsinfo.inquirer.net/419261/k-12-still-struggling. Accessed 14 March 2024.

Pernia, R. A. (2019). Human rights in a time of populism: Philippines under Rodrigo Duterte. *Asia-Pacific Social Science Review, 19*(3), 56–71.

Philippines places 158th out of 180 in environmental sustainability ranking. (2022, June 21). *BusinessWorld Online*. https://www.bworldonline.com/infographics/2022/06/21/456316/philippines-places-158th-out-of-180-in-environmental-sustainability-ranking/. Accessed 14 March 2024.

Pontifical Academy for Life. (2020). Rome call for AI ethics. *Romecall.org*. https://www.romecall.org/wp-content/uploads/2022/03/RomeCall_Paper_web.pdf. Accessed 14 March 2024.

Regilme, S. S. F., Jr. (2021). Visions of peace amidst a human rights crisis: War on drugs in Colombia and The Philippines. *Journal of Global Security Studies, 6*(2), 1–19. https://doi.org/10.1093/jogss/ogaa022

Roberts, H., Cowls, J., Hine, E., Mazzi, F., Tsamados, A., Taddeo, M., & Floridi, L. (2021). Achieving a 'Good AI Society': Comparing the Aims and progress of the EU and the US. *Science and Engineering Ethics, 27*(68), 1–25. https://doi.org/10.1007/s11948-021-00340-7

Rosales, M. A., Magsumbol, J. V., Palconit, M. G. B., Culaba, A. B., & Dadios, E. P. (2020). Artificial intelligence: The technology adoption and impact in The Philippines. In *2020 IEEE 12th International Conference on Humanoid, Nanotechnology, Information Technology, Communication and Control, Environment, and Management (HNICEM)* (pp. 1–6). IEEE. https://doi.org/10.1109/HNICEM51456.2020.9400025

Salac, R. A., & Kim, Y. S. (2016). A study on the internet connectivity in The Philippines. *Asia-Pacific Journal of Business Review, 1*(1), 67–88. https://doi.org/10.20522/APJBR.2016.1.1.67

Selvamani, M. J. P. (2022). Prepare the future: Rome call for AI ethics. *Journal of Data Analysis, 17*(4), 49–62. https://doi.org/10.6338/JDA.202212/SP_17(4).0005

Sevillano, S. (2022, April 4). DepEd, Microsoft note tech's role in inclusivity. *Philippine News Agency*. https://www.pna.gov.ph/articles/1171380 Accessed 14 March 2024.

Torneo, A. R., & Mojica, B. J. (2020). The strategic performance management system in selected Philippine National Government agencies: Assessment and policy recommendations. *Asian Politics & Policy, 12*(3), 432–454. https://doi.org/10.1111/aspp.12540

United Nations General Assembly. (1948). *The universal declaration of human rights*. United Nations General Assembly.

Varela, A. (2003). The culture perspective in organization theory: Relevance to Philippine public administration. In V. A. Bautista, M. C. P. Alfiler, D. R. Reyes, & P. D. Tapales (Eds.), *Introduction to public administration in The Philippines: A reader* (pp. 438–472). University of the Philippines.

World Bank. (2005). *Natural disaster risk management in The Philippines: Reducing vulnerability*. World Bank. https://openknowledge.worldbank.org/handle/10986/8754

Xu, W. (2019). Toward human-centered AI: A perspective from human-computer interaction. *Interactions, 26*(4), 42–46. https://doi.org/10.1145/3328485

Yeung, K. (2020). Recommendation of the council on Artificial Intelligence (OECD). *International Legal Materials, 59*(1), 27–34. https://doi.org/10.1017/ilm.2020.5

Zuiderveen Borgesius, F. J. (2020). Strengthening legal protection against discrimination by algorithms and artificial intelligence. *The International Journal of Human Rights, 24*(10), 1572–1593. https://doi.org/10.1080/13642987.2020.1743976

Chapter 6
Artificial Intelligence and Human Flourishing in the Everyday World: Reflections from Christian Public Theology

Stephen Garner

Abstract Artificial intelligence (AI) has transitioned from laboratory research to becoming an integral part of everyday life in technologies such as robotics, natural language processing, facial recognition, machine learning, autonomous transportation, and various planning and problem-solving systems. The practical implications of AI in our daily lives are often more pressing than philosophical debates about AI's potential to exhibit human-like qualities such as reason, creativity, consciousness, or emotional companionship. While these kinds of questions are important, the application of these AI technologies in the everyday world is too many far more important than their ontological nature.

The fusion of AI with so many different aspects of the everyday world, where it and other digital technologies become the very environment in which we live and breathe and have our being, raises ethical questions concerning how we use AI wisely to promote human flourishing and care of the wider world. This essay adopts a public theological perspective, incorporating both Christian insights and other viewpoints, to explore AI's role in big data, surveillance, and the workplace, aiming to support human well-being in the AI era.

Keywords Artificial intelligence · Public theology · Surveillance · Wellbeing · Databases · Labor · Social ethics

---

S. Garner (✉)
Laidlaw College, Auckland, New Zealand
e-mail: sgarner@laidlaw.ac.nz

L. Checketts, B. S. B. Chan (eds.), *Social and Ethical Considerations of AI in East Asia and Beyond*, Philosophy of Engineering and Technology 47,
https://doi.org/10.1007/978-3-031-77857-5_6

6.1 Introduction

Increasingly artificial intelligence technologies are moving from the laboratory into mainstream everyday life, permeating a huge range of human activities, seen and unseen. These technologies, including robotics, natural language processing, facial recognition, machine learning, autonomous transportation, and various planning and problem-solving systems, create a sense of wonder at the human creativity that produced them and a parallel anxiety that their use, indeed their very existence, might be detrimental to human flourishing. Part of that anxiety relates to concerns about whether artificial intelligence might or does possess attributes commonly associated with what is uniquely human, possess the religious concept of a 'soul', or demonstrate the capacity for emotional and physical companionship. While these are important concerns, for many people it is the role that artificial intelligence will play in the politics of the everyday that are more pressing matters, and it is to those matters that this paper pays attention.

We will do this using the lens of public theology, which is located in a constructive and robust dialog between religious faith, in this case, Christianity, and the wider world in order to promote public good and human flourishing. To begin we will sketch a little about artificial intelligence and then the nature of public theology, before turning to three areas—big data, surveillance, and the workplace—and locating ethical responses to them drawn, in part, from Christian theological reflection on human flourishing. Through this, we seek to offer something constructive and distinctive into artificial intelligence discussions to resource individual and community human flourishing in the everyday world.

6.2 Artificial Intelligence in Context

Given the range of places where artificial intelligence technologies turn up, some definitions are helpful for shaping the following discussions. One fairly typical definition of artificial intelligence provided by theologian computer scientist, Anne Foerst, is that "AI is the part of computer science and electrical engineering that attempts to build machines that are capable of performing functions for which we intuitively presume intelligence to be necessary" (2004, 66). While succinct, Foerst notes this introduces some immediate problems or ambiguities including how terms such as 'intuitive' and 'intelligence' are often defined by an individual's own worldview and their socio-cultural location.

Moreover, this definition limits artificial intelligence research and development to the disciplines of computer science and electrical engineering, and while those fields are important, other fields such as psychology, cognitive science, philosophy, and neuroscience also contribute, along with development and implementation being carried out in government, military, and commercial contexts. We might also ask how terms such as 'intelligence' and 'intuition' are imagined within the different cultures of China, Japan, and wider East Asia, and how those understandings might differ from those in the Americas, Europe, Africa, Oceania, and other parts of Asia.

One attempt to address these kinds of ambiguities is to broaden the scope somewhat beyond machines, as cyberneticist Kevin Warwick does when he states: "Artificial intelligence is not the study of computers, but of intelligence in thought and action. Computers are often its tools, because its theories are usually expressed as computer procedures that enable machines to do things that would require intelligence if done by people" (1998, 94). This helpfully highlights that beyond artificial intelligence's pragmatic developments, there are also aspects of artificial intelligence that are concerned with the desire to understand human intelligence and to replicate that using digital technologies. More recently, digital anthropologist, Beth Singler, encapsulates this in her definition of AI, where she says:

> Simply put, AI as a field involves working towards rational agents that respond to stimuli, perhaps in ways comparable to modes of human cognition. To this end, AI as a field can involve reasoning, planning, learning, natural language processing, sensory perception, emotional processing and, in robotic forms, the ability to move and manipulate objects (Singler, 2018, 217).

Singler's definition alerts us to the idea that artificial intelligence involves rationality and reasoning carried out by created agents in ways similar to how human beings do it and in ways that produce similar results but with different underlying mechanisms. It is this nuance, that Russell et al. (2010) develop in their four-fold typology of artificial intelligence systems, which groups artificial intelligence systems into: systems that think like humans; systems that think rationally, systems that act like humans; and systems that act rationally. Each of these approaches engenders different aspects of anxiety about artificial intelligence which feeds into public concerns, even in the limited contexts we will examine. However, before that, we shall consider what exact public theology is and why it might suggest particular courses of action in relation to artificial intelligence.

6.3 The Scope of Public Theology

Public theology is a form of theology that focuses primarily on the public good and welfare of society. As such, it draws on the faith traditions of a particular religious community and insights from disciplinary experts to offer something distinctly gospel or "good news" into public discourse (Forrester, 2004, 6). In doing this, it aims to speak persuasively, credibly, and intelligibly into various public spaces including government, business, local communities, the academy, and civil society. Stephen Pattison supplies one definition of this when he states public theology should:

> address issues of general public concern, in a genuinely public arena, in a publicly accessible way, using publicly comprehensible concepts and mechanisms. This with a view to effecting some kind of transformation of public views, policies and actions. Furthermore, it should be recognized as making a useful contribution by the public at large (Pattison, 2007, 212).

In this paper, I am approaching the public theological task through the lens of Christianity, including both Protestant and Roman Catholic traditions, and asking how those particular religious communities might engage with artificial intelligence, either in its totality or with a particular aspect, in order to offer something promoting a public good and wider human flourishing.

Communicating this theologically-informed reflection on matters raised by communities can be approached variously. We might take a "common-currency" approach, where some common ground with other groups is located through the use of a shared secular discourse: the God-language is muted, and there is the danger that the religious identity of a group can be lost, but it can prove a helpful way for disparate groups to work together. Alternatively, a religious community might seek to "out-narrate" competing narratives in the public square by offering a "distinctive discourse" rooted in the language of faith. Here, it is the lived agency of the religious community that shows to the wider world how the practice of faith and life promotes flourishing in the world (Marshall, 2005). Others, such as Robert Gascoigne, pick a *via media* in which religion is not introduced too early to the conversation to reduce the barrier to being heard, but religion should not be left out either, losing the identity of those speaking (Gascoigne, 2001). It is this latter approach that I tend to favour in my work, with the starting point being the concerns or matters raised in the wider community or *polis*, in this case regarding artificial intelligence, rather than in the church-facing concerns.

In passing, though, it is pertinent to note that there are various church-facing concerns regarding artificial intelligence technologies ranging across anxiety that human technological agency is crossing into areas reserved for God alone, that developments in artificial intelligence are applying pressure on religious understandings of being human and personhood, that generative artificial intelligence technologies will affect worship, liturgy, music, and preaching, and even in some cases that artificial intelligence is connected with imminent apocalyptic and eschatological events (Geraci, 2010).

6.4 Engaging with AI, Big Data, Surveillance, and Work

With those definitions now given, we turn now to three particular examples where artificial intelligence technologies are generating anxieties in the everyday world: the application of these technologies to vast repositories of data; the more specific application of the technologies in the area of surveillance; and concerns being raised about the impact of artificial intelligence on the workplace.

6.4.1 Big Data

One particular anxiety raised regarding artificial intelligence technologies is the application of those technologies to what is colloquially called "big data." Appearing towards the end of the 1990s, the term refers to "techniques for conceiving, designing, and developing vast amounts of information and operating systems that can gather, store, and process it" (Spector et al., 2023, 13). Of course, the actual definition of "big" changes over time as storage capacity increases and techniques for accessing and manipulating develop, but regardless of the size of the data repositories, questions emerge over the collection, storage, and use of all sorts of data.

Historically, Spector et al. argue that the first questions concerning "big data" asked primarily about what data should be collected, followed by deciding which algorithmic processes to use on that data. They note it was only later that questions emerged about the goals to be achieved by amassing of this data, followed by how to optimize access to it, and still later, the ethical questions of who is being served by these collections of data collection and whether their use is ethical and fair to various parties (Spector et al., 2023). Thus, the ethical questions emerge well after the technical ones. From a public theological view, the crucial question concerns how the collection and use of "big data" promotes human flourishing, typically maintaining the dignity and integrity of the persons whose data is captured, but also of those who create the technology and manage these systems.

Artificial intelligence technologies serve a key role in the administration and use of the sheer volume of data and information in these repositories, which are beyond the scope of individuals and groups of people to analyze and interpret this data without increasingly complex sets of tools. Systems can cluster and categorize data, detect and predict patterns in that data, create models of the real world from datasets, as well as optimize and reduce the entropy or complexity in those models to produce the smallest, most efficient models. The current interest in generative artificial intelligence systems based on large language models (LLMs), such as OpenAI's ChatGPT (Generative Pre-trained Transformer) and DALL-E systems, Google's BARD/Gemini, and the Midjourney system, depends upon training on very large datasets with equally large parameter sets (Pichai, 2023).

On the one hand, these generative systems, which are improving with each iteration, are wondrous demonstrations of human technological proclivity and provide what looks like some very useful tools for information retrieval, summarization, and text and image generation. On the other hand, they engender a deep sense of anxiety and apprehension among different communities from their apparent transgression into areas reserved for human beings, the perceived threat to individual and communal livelihoods, and the potential for the application of artificial intelligence technologies to the vast repositories of data in ways that lessen human freedom and which weaponize the data and technologies. Some of these concerns, including those connected to surveillance and workplaces, as well as others, generate ethical reflection that draws from secular and religious contexts to speak to these matters.

In considering a secular approach to ethics in relation to data science, Spector et al. take as their basis the well-established Belmont Report ethical model developed for biomedical research with human participants (Spector et al., 2023; National Commission for the Protection of Human Subjects of Biomedical and Behavioral Research, 1978). This model proposes three key principles that all must be held together in shaping an ethical activity involving individuals and communities: (1) respect for persons, seen particularly in recognizing the autonomy of individuals and communities in making decisions, and advocacy for those with diminished ability to exert that autonomy; (2) a call to beneficence, that goes beyond simply doing no harm to seek maximum benefit for those concerned, while managing immediate and long-term risks to those persons; and (3) justice, seen in the access to resources, the fair, rather than equal, distribution of risks and benefits, and that those who stand to benefit the most, bearing the most risk. Each of these principles interacts with the others, leading to negotiated decisions around which principles take priority in different situations.

Application to data science, they argue, might be worked out in managing genuinely informed consent in digital contexts by allowing good autonomous decision-making through access to information that is not overly simplified, but not overloading persons with huge amounts of information or complexity. Similarly, managing the risk/benefit balance should take into account potential unintended consequences of the use of the information system, as well as a commitment to addressing abuse and negative consequences of these systems, and finally, commitment to the ongoing assessment of data systems and their uses, including the training data used for them, and redressing harms caused. An example of the latter is the way in which data used to train systems can harm persons because inherent biases supplied in the training data, such as has been found in cases where facial recognition systems or judicial support systems have unfairly targeted particular ethnic communities in terms of profiling and sentencing activities (Morris, 2023; Chowdhury, 2020; Smith & Miller, 2022; Fleischer, 2020; Donohue, 2019; Hartmann & Wenzelburger, 2021).

There is significant commonality between the kind of secular ethical framework described here, and close kindred such as that proposed by Beauchamp and Childress (2001), with the approaches within Christianity for ethical engagement with technology. At one level, Jesus' commandments in the Christian Gospels to love one's neighbor as oneself (Matt 22:35–40; Mk 12:28–34; Lk 10:27a) and to love one's enemies (Matt 5:44; Lk 6:27–36) shapes a basic motivational foundation for engaging with the development and use of any technology, including artificial intelligence. So too, the imperative in the Hebrew Scriptures to do justice, love kindness, and walk humbly with God (Mic 6:8) also points to conduct connected to the just use of technology, fidelity and truthfulness, and a sense of our own human finitude in our technological endeavors. These foundational elements overlap with ethical frameworks such as the one used by Spector et al. and found in variations of autonomy, non-maleficence, beneficence, and justice.

For example, Lutheran theologian and bioethicist, Ted Peters, contends that beneficence must be a primary technological motivation. Humanity is being drawn forward towards an end, and morality changes or adapts under that pressure. Under this scenario, he argues, human agency is morally obliged to strive to make the world a better place through the use of technology, because "[t]he situation as it is does not necessarily describe how it ought to be" (Peters, 1997, 20–21). Therefore, people need to engage with technology not just informed by the ethical principle of non-maleficence, but also beneficence. If we can use technology to do good, then we are obliged to do so. If we do not, then we reject the potential God has given humanity for social transformation found in being bearers of God's image, Peters asserts (2004). In other words, a failure to act technologically for the benefit of others, leads to human beings failing in their mandate to be God's agents in the world.

Similarly, Roman Catholic biomedical ethics often takes as its starting point the criteria of autonomy, non-maleficence, beneficence, and justice but then adds additional layers of nuance shaped by the Catholic church's reflection on its faith. Thus, biomedical, and by extension, technological ethics, reflect concerns for the sacredness of life and the innate dignity of human beings, commitments to truth-telling and the common good for individuals and communities, as well as advocacy against the unjust use of technology and for those who are most vulnerable (Mackler, 2003).

These reflections of Christian technological ethics function in two directions. One, towards the various Christian individuals and communities in their own engagement with human technological agency in their own contexts, and a second direction, that of public theology towards wider society and the promotion of the common good. One example of this kind of public theological engagement in the context of artificial intelligence is seen in the 2020 "Rome Call for AI Ethics," a collaboration between the Roman Catholic Church, government, academies, and industry to establish ethical guidelines for artificial intelligence research and application.

> The Rome Call for AI Ethics is a document signed for the first time on February 28th 2020, in Rome, to promote an ethical approach to artificial intelligence. The idea behind it is to promote a sense of shared responsibility among international organizations, governments, institutions and technology companies in an effort to create a future in which digital innovation and technological progress grant man his centrality. Pointing to a new algorethics, the signatories committed to request the development of an artificial intelligence that serves every person and humanity as a whole; that respects the dignity of the human person, so that every individual can benefit from the advances of technology; and that does not have as its sole goal greater profit or the gradual replacement of people in the workplace (RenAIssance Foundation, 2023b).

The Rome Call document seeks to find, in the common currency approach of public theology, common ground for a set of principles to shape artificial intelligence use. Focusing upon the areas of ethics, education, and rights the document presents six key ethical principles to shape this technological agency:

1. Transparency: in principle, AI systems must be explainable
2. Inclusion: the needs of all human beings must be taken into consideration so that everyone can benefit and all individuals can be offered the best possible conditions to express themselves and develop

3. Responsibility: those who design and deploy the use of AI must proceed with responsibility and transparency
4. Impartiality: do not create or act according to bias, thus safeguarding fairness and human dignity
5. Reliability: AI systems must be able to work reliably
6. Security and privacy: AI systems must work securely and respect the privacy of users.

Recently, the Rome Call has expanded to include a broader engagement with other religious faiths in relation to artificial intelligence ethics. In early 2023, representatives from Jewish and Muslim communities joined with Catholic representatives under the banner of "AI Ethics: An Abrahamic Commitment to the Rome Call" to affirm the six principles and to issue the following joint statement (RenAIssance Foundation, 2023a):

We call upon all stakeholders, including but not limited to business, developers, governments, civil society, and academics to

- Uphold the algorethics principles, enshrining them in company policies and developing international regulation to maintain them;
- for those in authority or positions of power to refrain from the development of AI technologies for partisan, national or individual gain;
- for governments to refrain from the weaponization of AI technologies, which threatens the peaceful co-existence of us all.

The development of the neologism, algorethics (algorithm ethics), captures something of the desire for human freedom or autonomy to be preserved in the face of persons being reduced to data, something which is very much the case with the following concerns around surveillance.

6.4.2 *Surveillance*

Another anxiety that arises from the use of artificial intelligence systems with big data is the area of surveillance. Many of the principles mentioned in the previous section highlight a concern for the ethical and just use of digital data including elements of truth-telling, advocacy for those most vulnerable, and pursuit of beneficence as a primary motivation for the application of technology.

Surveillance is often portrayed as a singularly bad thing, visualized primarily in terms of systems for maintaining the societal status quo or privileges of an elite. In actuality, surveillance is something more ambiguous where, on the one hand, certain aspects of society require it to be done well to protect borders, restrict and prosecute crime, monitor financial and market irregularities, ensure supply chains function, manage traffic networks, safeguard the allocation of resources such as electricity and water, and provide safe and secure public spaces. Yet, on the other hand, there are genuine concerns about how much individual and community life is no longer private, and indeed threatened, by being permanently in the digital gaze.

The use of artificial intelligence systems for systems such as facial recognition was noted earlier, along with problems that can arise from that, but perhaps a more significant problem with respect to the collection of human data for analysis are the potential limitations of the data collected and how that limited data is used. Theologian and ethicist Eric Stoddart raises a number of significant concerns in this area, two of which will be briefly sketched here.

The first of these is Stoddart's concern with how digital technology intersects with Western democracy and reshapes that in both subtle and not-so-subtle ways. For example, the shift in Western contexts from governments delivering services directly to their citizens to one where those services are devolved to public and private agencies overseen by the government is tied, in part, to digital technologies being used to oversee and maintain the network of relations between citizens, companies, government agencies, civil society, and even other governments (Stoddart, 2008). This development puts significant personal and private data in the hands of both public and private agencies, each of which may be operating under different codes of practices governing the use of that data, as well as serving more than one interest.

Into this space, the ways artificial intelligence systems are used become myriad including the development of "chatbot" interfaces to manage and negotiate citizens engagement with these agencies, such as for health, social welfare, or business, to the use of artificial intelligence decision-making systems to process personal data, and the ability to link data between agencies and identity using artificial intelligence to detect patterns or anomalies, in spite of protections that might be in place legally to prevent that. Significant questions are raised about who actually has responsibility for how people treated within these systems and whether their citizenship is diminished by them. Stoddart notes that this is particularly the case for those who lack the information technology skills and infrastructure, falling on the wrong side of the "digital divide," which results not just in marginalization within society, but also a significant disincentive to participate as citizens in wider society (Stoddart, 2008, 343).

This capturing of personal data leads into the second problem that Stoddart envisages, that of the creation of digital "bodies" in these systems that represent their flesh-and-blood analogues. It is these digital bodies that serve the basis for the management of people and decision-making concerning them, but by necessity, these digital bodies can never capture the fullness of a person's being and lived experience. Stoddart describes these constructed bodies as:

> Data-doubles constructed from the informationized body (both above and beneath its surfaces) are key elements in contemporary surveillance systems. Whilst it might once have been legitimate to draw a clear distinction between virtual and real worlds this is now anything but obvious. The gaze here is far more than mere watching; entities akin to selves are constructed and acted upon in the "real" world. What is effected in cyberspace or virtual reality may well have consequences that are experienced in vivid material reality by people who are denied entitlement or otherwise marginalized politically, economically, or socially (Stoddart, 2019, 178–179).

It is these kinds of scenarios that ethical frameworks such as the Rome Call seek to engage with to set out what needs to be in place to maintain the dignity of the human persons and communities that are captured in these systems; to direct their trajectory towards something the promotes the common good and human flourishing, rather than reducing persons to digital entities to be operated on by increasingly impersonal systems. For all the recent debate about whether we should be developing ethical frameworks for how we treat artificially intelligent systems and whether or not those systems are becoming persons, we still have so much work to do concerning the treatment of human persons and communities in these digital contexts.

Stoddart's own response is to bring the issue of how people are represented to the forefront, and to work from a variety of perspectives, including theological, towards recognizing persons as presenting with a complex mix of embodiment, digitality, and human experience, to work against the misrepresentation of persons and communities, and to acknowledge that as persons we are both those surveilled and complicit with those who surveil (Stoddart, 2019).

6.4.3 Artificial Intelligence in the Workplace

The final area that to be discuss briefly concerns the challenges and opportunities posed by artificial intelligence technologies in the workplace. This stands in continuity with other historic changes that workplaces have negotiated in response to technological developments including mass production in the early twentieth century, developments in media and communications technologies, the use of personal computers, factory automation and industrial robotics, and more recently Internet-based and mobile technologies. In the contemporary context, the sheer rapid impact of artificial intelligence technologies for things like content creation, whether text or images, has caused significant anxiety across a range of professions, workplaces, and communities, perhaps more so than the topics of big data and surveillance noted previously.

The anxieties regarding things like Large Language Model (LLM) systems in relation to the workplace have, up until this point, I think, centered around two particular concerns. The first of these concerns the body of material that has been used to train the LLMs and whether that impinges on the copyright, intellectual and cultural property, and other artistic sensibilities of those whose profession is in the creative arts and industries. The second concern is that these systems might either be good enough or will be good enough to replace not just certain creative sectors, but also a significant tranche of run-of-the-mill, everyday jobs in the service sectors. Both of these concerns connect with a sense of human dignity being diminished in some way, and the very real potential for unemployment. For many this might be described, as Jacques Ellul described it well over 50 years ago, as a kind of technological determinism that seeks the most efficient means to an end with little regard for the human cost (Ellul, 1964).

Statements like the Rome Call offer principles to consider in relation to artificial intelligence and workplace. In particular, transparency, which suggests the artificial intelligence systems must be explainable speaks, in part, to the data being used to train systems and the data and algorithms being used to made decisions in the workplace. It also speaks to the veracity of 'texts' produced, to those responsible for that content and its effects, and identification as to whether a text has been produced by an artificial intelligence system, a human being, or a combination of both. The Call's principle of inclusion, which is connected to beneficence, asks whose end is served in the application of this technology, and whether it will promote a common good respecting the dignity and personhood of individuals and communities.

The Rome Call is clearly shaped by the body of Roman Catholic social teaching on the dignity of human persons, which includes theological reflection on the dignity of human work. This is linked specifically to an understanding that human beings have a subjective vocation to be agents in the world including human labor. For example, the Compendium of the Social Doctrine of the Church notes that:

> *This subjectivity gives to work its particular dignity, which does not allow that it be considered a simple commodity or an impersonal element of the apparatus for productivity.* Cut off from its lesser or greater objective value, work is an essential expression of the person, it is an "*actus personae*". Any form of materialism or economic tenet that tries to reduce the worker to being a mere instrument of production, a simple *labour force* with an exclusively material value, would end up hopelessly distorting the essence of work and stripping it of its most noble and basic human finality.
>
> *The human person is the measure of the dignity* of work: "In fact there is no doubt that human work has an ethical value of its own, which clearly and directly remains linked to the fact that the one who carries it out is a person" (Pontifical Council for Justice and Peace, 2004, III.a.271).

This Catholic theological understanding of human labor then shapes how, within the context of artificial intelligence in the workplace, a public theological endeavor might seek, working with others, to promote policy and environments that promote human personhood and dignity derived from work. This then engages with both the creative products of human labor being used by artificial intelligence systems, as well as how workplaces wisely and responsibly use those systems. It is, we might say, somewhat aspirational in the face of powerful economic and political forces, but it does offer something distinct into the discussion in the public square raised by anxieties concerning artificial intelligence.

6.5 Conclusion

This paper has sketched, we hope, how some of the broad outlines of public theology might engage with artificial intelligence as it is experienced in some different areas of the contemporary world. It seeks to put the objective of human flourishing and dignity, both for individuals and communities, as front and center of this discussion, and to highlight some of the particular concerns that are both being raised and engaged with through public theology. As such, we note the intersection of artificial intelligence with "big data", surveillance, and the workplace, and some of the ways that theologians and others within their religious communities are working with others to promote the wise use of technology for the service of all humanity.

There is a saying, attributed to J.H. Oldham, who developed what he called "middle axiom" prior to the Second World War, for the church and others to engage with social issues, that this kind of public engagement and theology is too important to leave just to the theologians. We need to have a diversity of voices from all walks of life in the room to have the robust conversations that need to be had, and I look forward to hearing some of those over the next few days.

References

Beauchamp, T. L., & Childress, J. F. (2001). *Principles of biomedical ethics* (5th ed.). Oxford University Press.

Chowdhury, H. (2020, June 10). Racial profiling concerns lead IBM to scrap facial recognition. *Daily Telegraph* (London).

Donohue, M. E. (2019). A replacement for Justitia's scales?: Machine learning's role in sentencing. *Harvard Journal of Law & Technology, 32*(2), 657–678.

Ellul, J. (1964). *The technological society* (J. Wilkinson, Trans.). Vintage Books.

Fleischer, R. S. (2020). Bias in, bias out: Why legislation placing requirements on the procurement of commercialized facial recognition technology must be passed to protect people of color. *Public Contract Law Journal, 50*(1), 63–89.

Foerst, A. (2004). *God in the machine: What robots teach us about humanity and god* (p. 2004). Dutton.

Forrester, D. B. (2004). The scope of public theology. *Studies in Christian Ethics, 17*(2), 5–19.

Gascoigne, R. (2001). *The public forum and christian ethics*. Cambridge University Press.

Geraci, R. M. (2010). *Apocalyptic AI visions of heaven in robotics, artificial intelligence, and virtual reality*. Oxford University Press.

Hartmann, K., & Wenzelburger, G. (2021). Uncertainty, risk and the use of algorithms in policy decisions: A case study on criminal justice in the USA. *Policy Sciences, 54*(2), 269–287. https://doi.org/10.1007/s11077-020-09414-y

Mackler, A. L. (2003). *Introduction to jewish and catholic bioethics: A comparative analysis*. Georgetown University Press.

Marshall, C. (2005). What language shall I borrow? The bilingual dilemma of public theology. *Stimulus, 13*(3), 11–18.

Morris, A. (2023). The dilemma of digital colonialism unmasking facial recognition technology and data sovereignty in Aotearoa New Zealand. *Policy Quarterly, 19*(1), 80–87. https://doi.org/10.26686/pq.v19i1.8109

National Commission for the Protection of Human Subjects of Biomedical and Behavioral Research. (1978). *The Belmont report: Ethical principles and guidelines for the protection of human subjects in research department of health, education, and welfare*. https://www.hhs.gov/ohrp/regulations-and-policy/belmont-report/index.html

Pattison, S. (2007). *The challenge of practical theology: Selected essays*. Jessica Kingsley.

Peters, T. (1997). Cloning shock: A theological reaction. In R. Cole-Turner (Ed.), *Human cloning: Religious responses* (pp. 12–24). Westminster John Knox Press.

Peters, T. (2004). Is our DNA sacred? *Response: The Seattle Pacific University Magazine* 26, no. 7. Accessed 29 April 2014. http://www.spu.edu/depts/uc/response/summer2k4/dna.html

Pichai, S. (2023). An important next step on our AI journey. *The Keyword* (Internet). Google Incorporated, 20 March 2023. https://blog.google/technology/ai/bard-google-ai-search-updates/

Pontificial Council for Justice and Peace. (2004). *Compendium of the Social Doctrine of the Church.* The Vatican (Vatican City). https://www.vatican.va/roman_curia/pontifical_councils/justpeace/documents/rc_pc_justpeace_doc_20060526_compendio-dott-soc_en.html

RenAIssance Foundation. (2023a). *AI ethics: An abrahamic commitment to the Rome call.* Updated 10 January, 2023, Accessed 29 March 2023. https://www.romecall.org/wp-content/uploads/2023/01/AI-Joint-Declaration-5-Jan1-1.pdf

RenAIssance Foundation. (2023b). *Rome call for AI ethics.* RenAIssance Foundation, Updated 23 March 2023, Accessed 20 March 2023. https://www.romecall.org/wp-content/uploads/2022/03/RomeCall_Paper_web.pdf

Russell, S. J., Norvig, P., & Davis, E. (2010). *Artificial intelligence: A modern approach* (Prentice Hall Series in Artificial Intelligence) (3rd ed.). Prentice Hall.

Singler, B. (2018). An introduction to artificial intelligence and religion for the religious studies scholar. *Implicit Religion, 20*(3), 215–231. https://doi.org/10.1558/imre.35901

Smith, M., & Miller, S. (2022). The ethical application of biometric facial recognition technology. *AI & SOCIETY, 37*(1), 167–175. https://doi.org/10.1007/s00146-021-01199-9

Spector, A. Z., Norvig, P., Wiggins, C., & Wing, J. M. (2023). *Data science in context: Foundations, challenges, opportunities.* Cambridge University Press.

Stoddart, E. (2008). Yes, no, cancel: Clicking our way to a public theology of cyberdemocracy. *International Journal of Public Theology, 2*, 328–353. https://doi.org/10.1163/156973208X316216

Stoddart, E. (2019). Reforming bodies under surveillance: An urgent task for theological education. *International Academy of Practical Theology. Conference Series, 1*, 176–183. https://doi.org/10.25785/iapt.cs.v1i0.73

Warwick, K. (1998). *In the mind of the machine: The breakthrough in artificial intelligence.* Arrow.

Chapter 7
Artificial Intelligence and Human Person: A Chinese Christian Perspective

Pan-chiu Lai

Abstract The development of Artificial Intelligence (AI) raises some important ethical and social issues of public concern, including (1) whether and in what sense AI can be regarded as person; and (2) how human person may live in a technological society controlled through or by AI. This study attempts to address some of these questions by exploring the Christian understanding of person, the Confucian understanding of human person, and the recent multi-disciplinary studies of human person. It aims at articulating a multi-dimensional understanding of human person and to explore its implications for the understanding of human predicament in technological society and the prospect of humanity.

Keywords Person · Artificial intelligence · Image of god · Confucianism · Tillich

7.1 Introduction

The development of Artificial Intelligence (AI) raises some important ethical, legal, and social issues of public concern. Many of these may make one reconsider the concept of "person" and the predicament of the human person in a technological society.

The first issue relates to the concept of "person" itself. Other than the question whether and to what extent AI is different from a human person, one may have to consider if AI can be regarded as a person from a legal point of view. For example, if a car "driven" by an AI system or say a robot is involved in a traffic accident, one may have to ask whether and how a robot can be regarded as a "legal person" and liable to legal responsibility. Can the law court put a robot into jail or pass a "death sentence" on a robot? Can a robot have "private property" to pay the fine or settle

P.-c. Lai (✉)
The Chinese University of Hong Kong, Shatin, Hong Kong SAR
e-mail: pclai@cuhk.edu.hk

L. Checketts, B. S. B. Chan (eds.), *Social and Ethical Considerations of AI in East Asia and Beyond*, Philosophy of Engineering and Technology 47,
https://doi.org/10.1007/978-3-031-77857-5_7

the insurance claim? If the robot is not a legal person, will the legal liability be placed on the designer of the AI system, the producer of the "robot," or some other parties involved? With the recent developments in "deep learning," which requires the interaction with the environments involving human beings, is it possible that one day an AI can be regarded as a third kind of "person" (other than natural person and legal person)?

The second issue relates to the technological control of human person. It is no secret that a totalitarian state can employ AI for the purpose of social or political control. Even in a liberal society, the algorithm involved in social media may influence or even manipulate the acquisition of information, economic activities, lifestyle, value or even worldview of individuals and thus the human society. One may thus wonder if these developments will threaten human freedom, turning human persons into robots controlled by AI or through AI by technological company or totalitarian dictatorship.

This study will explore (1) the relevance of the concept of person for legal, scientific, and technological issues, (2) the Confucian understanding of the human being, and (3) the contemporary debate in Christian theology concerning the body and the soul. Based on these explorations, it will sketch a multidimensional understanding of personhood and spell out its implications for facing the challenges to human person in technological society.

7.2 Image of God and Human Person

The relationship between technological development and the Christian understanding of human person seems to be rather complicated. On the one hand, the Western technological advancement was partially inspired and supported by an interpretation of the biblical story that the human being was made in the image of God and is expected to gain knowledge and power almost the same as God (Noble, 1999, especially vii, 3–6, 9–11). On the other hand, contemporary technological developments may also challenge the traditional Christian understanding of soul and immortality. For example, the development of AI may envision an "immortal mind" that can survive the successive destruction of hardware in a way comparable to but different from the immortality of human soul conventionally identified with the image of God (Noble, 1999, 143–200). It is notable that the Christian theological concept of "image of God" can play an important role in this kind of discussion, but the concept of "person", which is also very important in Christian theology, may facilitate better cross-cultural communication and the dialogue with other disciplines.

The concept of "person" (*persona*) plays a prominent role in the Christian theological discussion concerning human beings and has a long and complicated history in the Western intellectual tradition (see, e.g., Rolnick, 2007, 10–57). In the western philosophical and theological tradition, the most common definition of "person" is arguably the one articulated by Boethius (c.475/7–526?), namely an "individual substance of a rational nature". Modern scientific and technological developments raise some queries against this definition of personhood. For instance, if personhood is defined mainly by individual rationality, whether and to what extent are apes also persons? Whether and in what sense can machines with certain intelligence be recognised as persons? Is it possible for computers becoming or being recognized as persons?

This individualistic and rationalistic understanding of personhood is also challenged by social and psychological theories of personhood, especially social psychology. Under their influences or inspirations, some contemporary Christian theologians advocate for a relational or social understanding of personhood (McFadyen, 1990, especially 69–109). This relational or social understanding of personhood relates not only the Christian understanding of human persons (Shults, 2003, 217–242), but also the divine Persons (Grenz, 2001). In fact, the concept of "person" is involved in the traditional understanding of the divine Trinity in terms of "three persons in one substance." The contemporary renaissance of Trinitarian theology, though partially inspired by the relational or social understanding of person in some other academic disciplines, was brought forth mainly by the retrieval of the Christian doctrinal resources, especially patristic theology (Schwöbel, 2014). With this retrieval of patristic theology, some contemporary theologians advocate "person" as an ontological category, and even tend to take the understanding of divine Person as the basis for the understanding of human person, rather than the other way round (See: Schwöbel & Gunton, 1991).

The concept of "person" is deeply involved in the development of Christian theology and directly relevant to the discussion in other academic disciplines (Gregersen et al., 2000; Van Huyssteen, 2011; and Welker, 2014). Furthermore, comparable notions of "person" can be found in various cultures, including Greek, African, Chinese, and Indian (Carrithers et al., 1985). It is noteworthy that the concept of "person" is used even by scholars of Theravāda Buddhism, which is famous for its doctrine of Anattā (meaning literally no-self or not-self) and rejection of the existence of an eternal soul (Collins, 1982). After all, the concept of "person" is also used in the legal context and can provide a useful starting point or contact point for dialogue between theology and other academic disciplines on public issues.

7.3 Person in Legal, Scientific, and Technological Perspectives

The concept of "person" plays an important role in the public discussion concerning ethical and legal issues because the concept of "person" is often associated with rights and duties, and applicable not only to "natural persons" but also "legal persons."

The recent advancements in sciences, including particularly evolutionary biology and neuroscience, with their emphases on the importance of the body for personal identity, tend to assume a monistic understanding of the human person and refute body-soul dualism and the related concept of the immortal soul. As these issues are deeply related to Christian theology, especially the Christian understanding of the human being and salvation, these advancements have sparked off serious theological debates between dualism and monism (or physicalism) (Murphy, 2006). In fact, these scientific developments raise a wide range of philosophical issues related to human being, including human uniqueness and human freedom (Gazzaniga, 2011; Murphy & Brown, 2007). Given the complexity of these issues, it is not surprising to see the extensive inter-disciplinary dialogues among theology, philosophy, and natural sciences, especially psychology and neuroscience, related to the human person (Crisp et al., 2016; Russell et al., 1999; Jeeves & Brown, 2009).

The concept of "person" is also involved in the discussion concerning human beings and technology. It is interesting to note that Alan Turing (1912–1954), the father of computer science and AI, was well aware of the possible implications of AI for the understanding of the human being (Guo, 2015a, b). Apart from the legal question concerning whether an artificial intelligence can be regarded as a legal person, a more fundamental question concerns the boundary or border between human person and machine. In the world of AI, it is possible and rather easy to "transfer" or "migrate" some sorts of intelligence system, software, data, etc. from one machine to another. In other words, there is no "organic" integration or unity between hardware and software, and this separable relationship seems to be analogous to the position of body-soul dualism, which, usually based on a sharp contrast between immortal/spiritual soul and mortal/material body, identifies a human person with the soul instead of the body.

This dualistic position is at odd with the understanding of the human person supported by recent developments in neuroscience, which emphasize the "embodiment" of the human person and the organic integration or unity between body and mind (Watts, 2000, 49). According to this integrated view, the mind cannot be separated from the brain or even more broadly from biological embodiment (Waters, 2006, 65). As neuroscientist Michael S. Gazzaniga points out, when the idea of AI was first conceived in the 1950s, it was not modelled after the human brain. Instead, the basic approaches of AI are based on mathematical rules, especially algorithms. Moreover, underlying the various apparent differences between the human brain and digital computer, the most important difference remains "the fact that the brain is hooked up to a biological body" (Gazzaniga, 2008, 360–362; see further Fuchs, 2021, 13–48). However, with further developments in computer technology, especially the "fusion" of machine with human body, the dividing line between human and machine becomes increasingly blurred (Stock, 2002, 20–21).

It is notable that some pioneers of AI consider the development of AI as an extension of the most valued human capacity, that eventually can surpass and supplant human capabilities. Some even envision that instead of merely another advancement in technology, AI constitutes a turning point in human evolution (Noble, 1999, 157). It is because after this turning point the evolution of *homo sapiens* has become designed and directed mainly by human beings themselves. It is imaginable that in the future, human beings are increasingly living among AI and even "cyborgs" (organisms with both biological and technological parts). Furthermore, even human beings themselves may remain "fyborgs" (biological humans with certain machines implanted in the human bodies), their bodies may be increasingly constituted by carbon and silicon more than by flesh and blood (Stock, 2002, 24–31). This signifies a shift of vision from AI to A-Life (Artificial Life) (Noble, 1999, 165–171), which seems to be supported by advancement in genetic engineering that can provide "unprecedented choices" to humankind concerning their own future (Chapman, 1999, especially 207–256).

The application of technology to the human body seems to signify a human attempt to transcend human finitude, especially the spatial and temporal limits, and to achieve some sorts of immortality by transforming humans into posthumans (Waters, 2006, 60). While the vision of AI looks for the survival of humans by becoming software rather than hardware, genetic engineering envisions the enhancement and even perfection of human body (Waters, 2006, 64). In religious terminology, the former looks for immortality of the soul instead of the body, whereas the latter envisions bodily immortality without experiencing death. These technological developments raise serious questions for Christian theology. In addition to the question whether "immortality" can be achieved by human efforts and whether it is ethical to pursuit these kinds of "immortality," a more fundamental question remains their implications for the understanding of the human person (See Herzfeld, 2002). It is noticeable that these technological visions for human future seem to assume that as the bearers of the image of God, human beings, in analogy to the omniscient and omnipotent God who can create the world out of nothing, are "creators" or "co-creators" bestowed with unrestricted technological powers to change everything (including human beings themselves) and even invent something "new" or "create" something not yet created by God. Perhaps one may wonder if this "technological" vision is also a product of the Western culture and whether East Asian culture, especially Confucianism, can offer a better alternative.

7.4 Human Person in Confucianism

It is rather common to assume that Christianity and Confucianism hold divergent views on human nature—whereas Confucianism upholds the goodness of human nature, Christian theology insists on the sinfulness of human nature. This kind of stereotype might overlook some important themes and trends in Christian theology, especially the affirmation of the nihilistic nature of sin and the transformation of humanity through unity with Divinity (See: Lai, 2006, 2008). Furthermore, the Confucian understanding of the human being is not restricted to the discussion of human nature. In its discussion concerning the human person, Confucianism exhibits certain similarities with Christian theology. Confucianism advocates for a relational understanding of the human person, highlighting not only the importance of human social relationship to the formation of personhood, but also the constitutive role of the relationship with Heaven for the human person. Furthermore, Confucianism tends to reject the Cartesian soul-body dualism and tends to assume an "embodied" view of the human person supported by cognitive science and neuroscience. In fact, there are studies comparing Confucianism with cognitive science, social psychology, and even evolutionary psychology, including theories concerning reciprocal altruism and empathy altruism (See, e.g.: Munro, 2002; Tan, 2012; Seok, 2013).

It is noticeable that these interpretations of Confucianism are compatible with and can enrich the Christian understanding of personhood (Lai, 2016). Whereas Christianity affirms the doctrine of creation out of nothing (*creatio ex nihilo*) and the creation as a manifestation of the divine omniscience and omnipotence, Confucianism understands the world in analogous to the natural growth of organism and identifies the character of Heaven exhibited in the flourishing of human lives in terms of benevolence (仁, *ren*), which is also a virtue endowed to human beings. For Confucianism, especially Neo-Confucianism, the human person participates not only in the divine virtue but also in the world in terms of "forming one body with ten thousand things of heaven and earth." The human person is conferred with the divine mandate of becoming assistants to the divine activities aiming at flourishing lives in a harmonious world. It is thus possible to articulate a Confucian interpretation of personhood in terms of virtue, especially benevolence, rather than power or intelligence. This interpretation may negatively reject an unlimited authorisation of human exploitation of nature, and positively offers important hints or clues for the development of ecological theology (Lai, 1998).

The overall Confucian understanding of "embodied" personhood may imply a multidimensional understanding of the human person, which includes the biological, psychological, social, cosmic, and even spiritual dimensions. This multidimensional understanding of the human person can resist the tendency of reducing the human person to "information," "software," or "intelligence" in the age of digital technology. It may also raise questions concerning the human future.

It is rather well-known that Confucianism takes the establishment of perpetual peace on earth, instead of escape from this mundane world, as the ultimate aim of one's personal life. It considers immortality in terms of "three immortalities" (三不朽), namely establishing virtue (立德), merit (立功), and word (立言). The crude fourfold typology for the visions of immortality driving the development of civilization, namely (1) immortality of soul, (2) living forever, (3) resurrection, and (4) legacy (Cave, 2012), may find their respective expressions in the vision associated with AI, the vision of genetic engineering, Christianity, and Confucianism, which adopts a sceptical or agnostic attitude towards questions about life after death. In line with this attitude, Confucianism tends to emphasise the enhancement of the well-being of the whole society rather than the survival of the individual life after death. Confucianism prefers to ground our understanding of the human person on actual experiences, including interactions with other human persons, in the present world, rather than speculations about life after death. So, Confucianism may query not only if the Christian vision is too speculative and remote from daily life, but also whether the other two visions are too individualistic or selfish. It would prefer to explore how technological developments may benefit the sustainable development of the whole world rather than the immortality of few individuals.

7.5 Contemporary Theological Approaches to Human Person

In response to the modern scientific and philosophical challenges to the traditional understanding of soul and the associated dualistic view of humanity, some contemporary theologians reconsider the controversial issues concerning monism and dualism. For example, John W. Cooper concludes his study of the biblical view(s) of humanity with a vindication of "Holistic Dualism" which affirms that the Biblical view of human nature is both holistic and dualistic. On the one hand, the Bible "emphasizes the religious, phenomenological, and functional integration of life;" on the other hand, it also asserts that "persons are held in existence without fleshly bodies until the resurrection." Cooper argues that while this "Holistic Dualism" is not absolutely incompatible with the modern scientific and philosophical understandings of the human being, the "monisms of materialism and ontological holism" cannot account for the intermediate state between death and resurrection and the personal identity at the resurrection (John Cooper, 2000, especially 230–231). Though Cooper might attempt to maintain a sort of balance between Dualism and Monism, as Joel B. Green observes, his presentation in the book leans more towards dualism but shifts later towards monism and argues that "the soul is neither a substance nor an entity" (Green, 2008, 9–10). Cooper's survey indicates that the backbone for the body-soul dualism concerns the question of life after death. However, the Christian understanding of life after death does not necessarily imply a body-soul dualism. According to the biblical accounts of resurrection, the resurrected Jesus Christ was not without body, and there seemed to be certain personal identity or continuity between the dead and the resurrected person.

In his study of the biblical view of humanity in dialogue with the natural sciences, Green makes the following points concerning continuity of personal identity in resurrection: (1) "our identity is formed and found in self-conscious *relationality* with its neural correlates and embodied *narrativity* or formative histories;" (2) death does not mean merely the biological cessation of one's body, but also socially the severance of all relationships as well as the fading of personal narrative; and (3) life-after-death requires embodiment or re-embodiment. Based on this understanding, "the capacity for resurrection, for transformed existence, is not a propriety intrinsic to the human person (nor to the created cosmos)." This interpretation of resurrection implies that the preservation of personhood in resurrection is constituted by relationality instead of individuality or rationality (Green, 2008, 179–180, emphasis original). Green ends his discussion with the following remarks:

> This reminds us, again, that the capacity for "afterlife" is not a property of humanity, but is a divine gift, divinely enacted. It also underscores the reality that, in eschatological salvation, we are not rescued from the cosmos in resurrection, but transformed with it in new creation (Green, 2008, 180).

Green's remark is reminiscent of not only Tillich's interpretation of hope for the eschatological participation of all lives in eternity (Lai, 1996), but also the Biblical text concerning the "new creation" in Jesus Christ (2 Cor. 5:17), which was reinterpreted philosophically by Tillich in terms of the "new being" with elaboration on its universal significance in conjunction with the doctrine of resurrection (Tillich, 1978a, 150–165). In other words, theological anthropology concerns not only eschatology but also Christology (understanding the person and nature of Christ). Apart from the eschatological approach of viewing personal identity through the perspective of bodily resurrection (see also: Peters et al., 2002, especially 89–182), it is also entirely legitimate and even desirable to adopt a Christological approach to theological anthropology (Gortez, 2017, 15–26).

From a theological perspective, whereas Jesus Christ as a person is recognised as the true *imago Dei* (Col. 1:15), other human beings are merely created after the image and likeness of God or the secondary images imitating the Image of God *per se*, namely Jesus Christ. In other words, Jesus Christ should be considered as an exemplary model for all human beings, and this concerns mainly the relationship between Jesus Christ and God (or more specifically God the Father). Different from Confucianism, which emphasizes the divine-human unity and tends to downplay the difference, Christianity tends to highlight the difference between humanity and divinity. This difference is exhibited not only in the doctrine of creation, which assumes the fundamental distinction between the Creator and creatures but also in Christology, especially the Chalcedonian formula, which assumes that all human persons share one human nature, while all divine persons share one divine nature. Following the insight of Chalcedon, the Byzantine theological tradition developed a distinction between *energeia* (activity/actuality) and *ousia* (nature/substance) in order to explain the unity between humanity and divinity in the deified person that while humanity shares the divinity in terms of *energeia*, but not *ousia*—not to say *hypostasis*, which is different even among the three divine persons. As Jesus Christ, who is the image of God *per se*, is the paradigm for deification, the deification of other human persons is basically a matter of "Christification" (Nellas, 1987, 121–139; Jordan Cooper, 2014). A corollary of this Christological understanding of deification is that the divine image embodied in human person is concerned primarily with human virtue, morality, humility, spirituality, benevolence, etc., instead of the divine omniscience or omnipotence. It is because Jesus Christ as the divine incarnation was characterized typically by "kenosis," self-giving, love, suffering, and reciprocity with other human beings, rather than clinging on to his own divine power or authority. This kind of kenotic love can be found not only in Christ's life and the divine creation but also in human altruistic behaviours. In other words, kenosis can be both a cosmological and an anthropological concept open to the dialogue between theology and sciences, including cosmology and evolutionary biology (Ellis, 2001; Jeeves, 2001). One may thus find that a Christological interpretation of the divine image diverges radically from the technological interpretation, and comes closer to Confucianism, which highlights Heaven's benevolence embodied in the life process as well as the human capacity for benevolence and altruism.

Regarding the eschatological perspective, especially the doctrine of resurrection, the biblical text in 1 Corinthians 15, the *locus classicus* on this subject, is full of pneumatological (i.e., relating to the Holy Spirit) terminology, and compares the "first Adam" as a "living soul" with the "last Adam" as "life-giving Spirit" (1 Cor. 15:45). In fact, in Romans 1:4, the resurrection of Jesus Christ is also ascribed to the work of the Holy Spirit. Borrowing the expression in the Nicene Creed (CE 381), as the "giver of life," the Holy Spirit gives not only the resurrected life but also the earthly human life as a divine gift. In this sense, the human being is "living on borrowed breath"—with the understanding that "breath" refers to *ruach* in Hebrew (breath/spirit), which corresponds to *pneuma* in Greek (Kelsey, 2009, 159–438). In other words, pneumatology, which is not to be restricted to the doctrine of the Holy Spirit, can provide a framework to account for a human person before and after resurrection. It is then unnecessary to affirm two contradictory views for human person—a holistic view for the human life before death and a dualistic view for the human life after death. It is thus possible and even desirable to explore a pneumatological approach to anthropology (Welker, 2021). As we are going to see, a pneumatological approach to anthropology is restricted neither to the discussion concerning human spirit as an entity apart from human body, nor the spiritual dimension of human person; rather, it can cover the whole multidimensional human life as spirit. Furthermore, the pneumatological and Christological approaches are compatible in their affirmation of the multidimensionality of human person, involving the physical, biological, psychological, and spiritual dimensions.

7.6 Multi-Dimensional Personhood and Technological Society

In response to the studies of personhood by other academic disciplines, some Christian theologians argue for a multidimensional understanding of personhood. For example, in his theological account of the human condition, Edward Farley (1929–2014) engages in dialogue with the relevant academic disciplines, and identifies three spheres of human reality, namely interhuman, social, and personal spheres, including the biological aspects and elemental passions of personal being. These characterisations of human beings imply that the human person is multidimensional, covering biological, psychological, and social dimensions (Farley, 1990). In a similar vein, Fraser Watts employs the term "the multifaceted nature of human personhood" to argue for a balanced and multifaceted approach to human person as a biocultural reality, and against the extreme positions of the cultural/sociological interpretation overwhelmed by biological one or vice versa (Watts, 2000). More explicitly, Veli-Matti Kärkkäinen advocates a "multidimensional monism" primarily in response to the debate concerning monism and dualism inspired by modern science (Kärkkäinen, 2016). He further elaborates this concept in his attempt to articulate a Christian systematic theology through comparison with other religions (Kärkkäinen, 2015, 306–349). It is rather unfortunate that Kärkkäinen's comparative studies omit the Chinese tradition, especially Confucianism, but as Kärkkäinen acknowledges, this multidimensional view of human person is pioneered by Paul Tillich (1886–1965) (Kärkkäinen, 2016, 215).

Unlike Kärkkäinen's unelaborated citation, Philip Hefner makes extensive references to Tillich, when engaging in the science-theology dialogue on personhood and *imago Dei.* Hefner accepts both the physicalist assumption as well as the evolutionary-emergent assumption. While the former assumes that the person is rooted in the brain and its body, in the latter the person has emerged within a set of evolutionary processes and is itself a process of emergence, which refers to the emergence of something new of a higher order of complexity from physical-biological processes (Hefner, 2000, 74–76). Hefner admits that his basic point of view is adopted from Tillich's concept of the multidimensional unity of life, which further implies that personhood is not only the creation of the Spirit but also a construction at the physical, biological, and cultural levels (Hefner, 2000, 76–77). Hefner further elaborates Tillich's view that on the one hand, personhood as spiritual individuality requires a centred-self, but on the other hand, it is defined by relations, involving cognitive and moral processes (Hefner, 2000, 80–84).

Regarding Tillich's understandings of life as a multidimensional unity and technological society, there are a few observations to be made. First, Tillich deliberately uses the term "dimension" instead of "level" because the metaphor of "level" may imply a sort of hierarchy or mutual independence among the levels (Tillich, 1978b, 12–15). This appears to be quite different from the emphasis on the concepts of "level" and "hierarchy" in the emergentist hypothesis, which is an important topic for the dialogue between theology and science (Peacocke, 2006; Deacon, 2007). Tillich's view is not against the theory of evolution or the emergentist hypothesis. It is noteworthy that when Philip Clayton elaborates the soteriology (related to salvation) of his "emergentist panentheism," which assumes an evolution-physicalist position, he refers to Tillich (Clayton, 2007, 340–341; also Clayton, 2008, 112–113). This indicates that Tillich's theology, especially his pneumatology, can be compatible with the evolutionary theory and emergentist hypothesis (Kung, 2020). In fact, the metaphor of "dimension" and the resultant rejection of hierarchy may have positive implications for science-theology dialogue. One of these implications is to refute the possible hidden assumption that some levels or dimensions should be "higher" or more valuable than the others, particularly the "dualistic" or "spiritualistic" view that emphasizes the spirit (or "spiritual") at the expense of the body (or the "physical"). Regarding its implications for the understanding of the mind-body problem, as Tillich himself succinctly summarises,

> In the moment in which we use the metaphor of "dimension and not the metaphor of "level" or "stratum," we can say that all dimensions are present in man[sic]. In the organic, the spiritual is present. In the spiritual, the inorganic is present, and both in the organic. I could say with most of the progressive biologists and neurologists that man's [sic] spiritual dimension is present in every cell of his body. This is, of course, a monistic interpretation of man[sic], which also implies the opposite, that in every spiritual act of man[sic], the function of every cell is noticeable. You cannot separate them. They are one and the same reality (Tillich, 1988, 115).

Tillich's reference to "the progressive biologists and neurologists" indicates his awareness of the possible relevance of his multidimensional understanding of life to natural sciences.

Second, based on Tillich's understanding of life as a multidimensional unity, the human person, who actualises and unites all dimensions of life, is a multidimensional unity with many dimensions, including biological, psychological, and spiritual ones, which are inseparable from each other. This clearly affirms the relevance of various academic disciplines to each other in their studies on the human person. Furthermore, according to Tillich, there should be no conflict or mutual interference among these dimensions (Tillich, 1978b, 15). On the one hand, it may further imply that the human religiosity or spiritual quest may have its biological and psychological bases. The seemingly "secular" biological or psychological studies of human beings, in principle at least, can be relevant to theology. On the other hand, a multidimensional view of personhood may also imply that personhood is not reducible to any one of the dimensions, neither the physical nor the spiritual. This multidimensional understanding of personhood may provide a more comprehensive understanding of personhood to counter the reductionist understanding advocated by atheistic materialism and facilitate interdisciplinary dialogue on the human person with proper recognition of the integrity and autonomy of the respective disciplines involved.

Third, as a philosopher closely related to the Frankfurt School, it is rather natural for Tillich to address the issues related to technology from time to time. In fact, Tillich reflected critically on various problems about technology, including not only the exploitations of natural environment and the threat of atomic bombs but also the impacts on human person, such as the "thingification," objectification, or dehumanisation in technological society (Tillich, 1988). Tillich criticised, for example, the technological tendency of emphasising conformity at the expense of creativity, not only in totalitarian regimes but also in Western technical society, where all the cultural contents might lose their creative powers and become matters of entertainment, sensation, sentimentality, etc. (Tillich, 1988, 134–135). Tillich was also well aware of the socio-political implications of technological advancements, particularly the growth of a scientific elite. He particularly warned, "Space exploration in the democratic world strengthens the antidemocratic elements, which are present in every democratic structure" (Tillich, 1988, 194). For Tillich, the elitism involved in the technological development, together with the elitist hope for the human future, which may include only a small number of the elites and exclude the majority of humankind, should be criticised and rejected by the eschatological universal salvation supported by the multidimensional understanding of life and personhood. Many of Tillich's observations or comments about technical society may perhaps remain valid in this age of digital technology.

Fourth, what Tillich aims to offer is not merely a diagnosis of the problems of human persons in technical society, but also a possible approach to deal with the problem. His philosophy of life, especially his understanding of the dimensions of life, may provide some sorts of hope in this kind of situation. He writes, "Although everything in the world can become a logical object, nothing in nature, not even an atom, is only or merely an existential object. Everything in nature shows resistance against becoming a mere object" (Tillich, 1988, 113). Tillich further suggests that Christianity can play a positive role in the resistance to the "de-humanisation" in technical society. In his own words,

> Christian action must point to the ultimate roots of personal being. It must show that man can maintain his nature and dignity as a person only by a personal encounter with the ground of everything personal.... It is the New Reality that is manifest in Christ and against which even technical society and its power of destroying the person as person cannot prevail (Tillich, 1988, 137).

This may mean that in its dialogue with sciences on human person in a technological age, Christian theology can and should on the one hand make references to biology and neuroscience to uphold a multidimensional understanding of life as well as personhood, and on the other hand explain how Jesus Christ and the Christian community may resist de-humanisation through affirming personhood. Regarding the former part, it is interesting to note that there are some biological and neuroscientific studies indicating the "transcendence" of the human person or human spirit in the sense that it is a biological imperative driving human life into ever-greater realms of being (Pearce, 2004). Furthermore, scholars from some other disciplines also find "transcendence" a facet of the human person. For example, from an anthropological or ethnographical perspective, the human being is a questing animal with "self-transcendent questing impulse" (Torrance, 1994, xiv, 50–57). This "spiritual" quest for transcendence can be found in science as well as religion, particularly myth and ritual, and there are biological, psychological, and linguistic foundations of the quest (Torrance, 1994, 18–49). Historically speaking, religions, particularly their rituals, have played an important role in human evolution (Bellah, 2011, 567). Given the social nature of ritual, these studies of human evolution and even "transcendence" of human spirit converge at the affirmation of the role of human community in the establishment of the human person. In other words, even the human spirit is also in a sense a social construction (Teske, 2000).

This affirmation of the constitutive role of the human community or personal relationship in the formation of the human person may cast light on the question about the future of humanity. If the human community is shaped in history, by the human persons involved, and in interaction with the environment, the human future is not determined entirely by technological advancement. Even if one can have a perfect human body or take up another body easily, there is no absolute guarantee of eternal life or immortality because life remains reliant on support from other lives. As the sci-fi movie "Voyagers" (2021) indicates, the human future relies not entirely on technology (such as genetic engineering and spacecraft) but also on interpersonal relationships. Without a proper relationship among the human persons, there will be suspicion, hostility, hatred, etc., and human beings may destroy each other with the most advanced technology available. Put it in a more Chinese way, perhaps for the future of humanity, the most vital is not the technological development of genetic engineering or AI (artificial intelligence), but the practice of another kind of "AI", which should be pronounced together as "ài", meaning "love"

(愛) in Chinese. From an evolutionary perspective, the future of humanity is not determined entirely by biological evolution and technological advancement; rather, it is also shaped by the evolution of culture, morality, religion, altruism, etc. (Hefner, 1993; Keogh, 2015).

7.7 Concluding Remarks

The above discussion is merely a preliminary thought-experiment rather than a systematic theological attempt. It aims to explore how East Asian Christians may make use of their cultural and religious resources as well as the findings from other academic disciplines to articulate a multi-dimensional understanding of the human person in response to the challenges constituted by contemporary technology, including technological social control through or by AI. With this interdisciplinary and inter-cultural/inter-religious approach, East Asian Christians may address some other issues of public concern beyond East Asia.

Acknowledgement This paper is a substantially abridged version of my article "Multi-dimensional Personhood in Sino-Christian Perspective" forthcoming in *Sino-Christian Studies* (2025).

References

Bellah, R. N. (2011). *Religion in human evolution: From the Paleolithic to the axial age*. The Belknap Press of Harvard University Press.

Carrithers, M., et al. (Eds.). (1985). *The category of the person: Anthropology, philosophy, history*. Cambridge University Press.

Cave, S. (2012). *Immortality: The quest to live forever and how it drives civilization*. Crown Publishers.

Chapman, A. R. (1999). *Unprecedented choices: Religious ethics at the frontiers of genetic science*. Fortress Press.

Clayton, P. (2007). Toward a constructive Christian theology of emergence. In N. Murphy & W. R. Stoeger (Eds.), *Evolution and emergence systems, organisms, persons* (pp. 325–344). Oxford University Press.

Clayton, P. (2008). *Adventures in the Spirit: God, world, divine action*. Fortress Press.

Collins, S. (1982). *Selfless persons: Imagery and thought in Theravāda Buddhism*. Cambridge University Press.

Cooper, J. W. (2000). *Body, soul and life everlasting: Biblical anthropology and the monism-dualism debate* (rev. ed.). William B. Eerdmans.

Cooper, J. (2014). *Christification: A Lutheran approach to Theosis*. Wipf & Stock.

Crisp, T. M., et al. (Eds.). (2016). *Neuroscience and the soul: The human person in philosophy, science, and theology*. William B. Eerdmans.

Deacon, T. W. (2007). Three levels of emergent phenomena. In N. Murphy & W. R. Stoeger (Eds.), *Evolution & emergence: Systems, organisms, persons* (pp. 88–110). Oxford University Press.

Ellis, G. F. R. (2001). Kenosis as a unifying theme for life and cosmology. In J. Polkinghorne (Ed.), *The work of love: Creation as kenosis* (pp. 107–126). Grand Rapids, MI.

Farley, E. (1990). *Good and evil: Interpreting a human condition*. Augsburg Fortress.
Fuchs, T. (2021). *In defense of the human being: Foundational questions of an embodied anthropology*. Oxford University Press.
Gazzaniga, M. S. (2008). *Human: The science behind what makes your brain unique*. HarperCollins.
Gazzaniga, M. S. (2011). *Who's in charge? Free will and the science of the brain*. HarperCollins.
Gortez, M. (2017). The madness in our method: Christology as the necessary starting point for theological anthropology. In J. R. Farris & C. Taliaferro (Eds.), *The Ashgate research companion to theological anthropology* (pp. 15–26). Routledge.
Green, J. B. (2008). *Body, soul, and human life: The nature of humanity in the bible*. Paternoster.
Gregersen, N. H., et al. (Eds.). (2000). *The human person in science and theology*. T&T Clark.
Grenz, S. J. (2001). *The social god and the relational self: A Trinitarian theology of the image Dei*. Westminster John Knox Press.
Guo, T. (2015a). Alan Turing: Artificial intelligence as human self-knowledge. *Anthropology Today, 31*(6), 3–7.
Guo, T. (2015b). Spirituality as re-conceptualisation of the self: Alan Turing and his pioneering ideas on artificial intelligence. *Culture and Religion: An Interdisciplinary Journal, 16*(3), 260–290.
Hefner, P. (1993). *The human factor: Evolution, culture, and religion*. Fortress Press.
Hefner, P. (2000). Imago Dei: The possibility and necessity of the human person. In N. H. Gregersen et al. (Eds.), *The human person in science and theology* (pp. 73–94). T & T Clark.
Herzfeld, N. L. (2002). *In our image: Artificial intelligence and the human spirit*. Fortress Press.
Jeeves, M. (2001). The nature of persons and the emergence of kenotic behavior. In J. Polkinghorne (Ed.), *The work of love: Creation as kenosis* (pp. 66–89). William B. Eerdmans.
Jeeves, M., & Brown, W. S. (2009). *Neuroscience, psychology, and religion: Illusions, delusions, and realities bout human nature*. Templeton Foundation Press.
Kärkkäinen, V.-M. (2015). *A constructive Christian theology for the pluralistic world, vol. 3, creation and humanity*. William B. Eerdmans.
Kärkkäinen, V.-M. (2016). 'Multidimensional monism': A constructive theological proposal for the nature of human nature. In T. M. Crisp et al. (Eds.), *Neuroscience and the soul: The human person in philosophy, science, and theology* (pp. 201–227). Grand Rapids, MI.
Kelsey, D. H. (2009). *Eccentric existence: A theological anthropology* (Vol. 1). Westminster John Knox Press.
Keogh, G. (2015). *The evolution of hope: Theological metaethics in the context of evolution*. Peeters.
Kung, W.-h. (2020). *Pneumatology and emergence: Paul Tillich in dialogue with science. PhD thesis*. The Chinese University of Hong Kong.
Lai, P.-c. (1996). A contextual reflection on Tillich's interpretation of Hope. *Ching Feng, 39*(4), 287–306.
Lai, P.-c. (1998). Christian ecological theology in dialogue with Confucianism. *Ching Feng, 41*(3–4), 309–344.
Lai, P.-c. (2006). Barth's doctrines of sin and humanity in Buddhist perspective. *Studies in Interreligious Studies, 16*(1), 41–58.
Lai, P.-c. (2008). Christian transformation of Greek humanism and its implications for Christian-Confucian dialogue. *Korea Journal of Systematic Theology, 22*, 245–269.
Lai, P.-c. (2016). Confucian understanding of humanity and rationality in conversation: A Chinese Christian perspective. In M. Mühling (Ed.), *Rationality in conversation: Philosophical and theological perspectives* (pp. 321–335). Evangelische Verlagsanstalt.
McFadyen, A. I. (1990). *The call to personhood: A Christian theory of the individual in social relationships*. Cambridge University Press.
Munro, D. J. (2002). Reciprocal altruism and the biological basis of ethics in Neo-Confucianism. *Dao: A Journal of Comparative Philosophy, 1*(2), 131–141.
Murphy, N. (2006). *Bodies and souls, or spirited bodies?* Cambridge University Press.

Murphy, N., & Brown, W. S. (2007). *Did my neuros make me do it? Philosophical and neurobiological perspectives on moral responsibility and free will*. Oxford University Press.

Nellas, P. (1987). *Deification in Christ: The nature of the human person* (N. Russell, Trans.). St Vladimir's Seminary Press.

Noble, D. F. (1999). *The religion of technology: The divinity of man and the spirit of invention [1997]* (Vol. 23). Penguin Books.

Peacocke, A. (2006). Emergence, mind, and divine action: The hierarchy of the sciences in relation to the human mind-brain-body. In P. Clayton & P. Davies (Eds.), *The re-emergence of emergence: The emergentist hypothesis from science to religion* (pp. 257–278). Oxford University Press.

Pearce, J. C. (2004). *The biology of transcendence: A blueprint of the human spirit*. Park Street Press.

Peters, T., et al. (Eds.). (2002). *Resurrection: Theological and scientific assessments*. William B. Eerdmans.

Rolnick, P. A. (2007). *Persons, grace, and god*. William B. Eerdmans.

Russell, R. J., et al. (Eds.). (1999). *Neuroscience and the person: Scientific perspectives on divine action*. Vatican Observatory Publications.

Schwöbel, C. (2014). Where do we stand in Trinitarian theology? In C. Chalamet & M. Vial (Eds.), *Recent developments in Trinitarian theology* (pp. 9–71). Fortress Press.

Schwöbel, C., & Gunton, C. (Eds.). (1991). *Persons, divine and human*. T&T Clark.

Seok, B. (2013). *Embodied moral psychology and Confucian philosophy*. Lexington Books.

Shults, F. L. R. (2003). *Reforming theological anthropology: After the philosophical turn to relationality*. Wm. B. Eerdmans.

Stock, G. (2002). *Redesigning humans: Our inevitable genetic future*. Houghton Mifflin.

Tan, M. (2012). An evaluation of Confucian sympathy from the perspective of evolutionary ethics. *Berliner China: Hefte / Chinese History and Society, 41*, 32–42.

Teske, J. A. (2000). The social construction of the human spirit. In N. H. Gregersen et al. (Eds.), *The human person in science and theology* (pp. 189–211). T&T Clark.

Tillich, P. (1978a). *Systematic theology* (Vol. 2. [1957]). SCM Press.

Tillich, P. (1978b). *Systematic theology* (Vol. 3. [1963]). SCM Press.

Tillich, P. (1988). In J. Mark Thomas (Ed.), *The spiritual situation in our technical society*. Mercer University Press.

Torrance, R. M. (1994). *The spiritual quest: Transcendence in myth, religion, and science*. University of California Press.

Van Huyssteen, J. W., & Wiebe E. P. (Eds.). (2011). *In search of self: Interdisciplinary perspectives on personhood*. William B. Eerdmans.

Waters, B. (2006). *From human to Posthuman: Christian theology and technology in a postmodern world*. Ashgate.

Watts, F. (2000). The multifaceted nature of human personhood: Psychological and theological perspectives. In N. H. Gregersen et al. (Eds.), *The human person in science and theology* (pp. 41–63). T&T Clark.

Welker, M. (Ed.). (2014). *The depth of the human person: A multidisciplinary approach*. William B. Eerdmans.

Welker, M. (2021). *In God's image: An anthropology of the spirit* (D. Stott, Trans.). William B. Eerdmans.

Part III
Comparative Theoretical Ethical Approaches

Chapter 8
AI and Information World-Making Ethics

Tianen Wang

Abstract Humans have been constructing an increasingly intricate layer of information atop the material world. With the advancement of artificial intelligence (AI), this information layer is evolving into a realm of heightened intelligence, effectively transforming into an information-centric domain for humanity. The advent of big data has paved the way for unprecedented progress in human endeavors, while the rise of AI signals a shift towards a world where human information activities increasingly shape reality. Consequently, novel ethical issues arise, centering on the ethics of shaping this information-based world. Thus, a new ethical framework, termed "information world-making ethics," has emerged. Within the dimension of information world-making ethics, the integration of rules and laws offers insight into the future trajectory of AI development and human-machine interactions. The fundamental challenge of informaiton world-making ethics lays in the ethics of foundational principles, with integrity, humankind's collective characteristic, and commonality being paramount. As AI progresses, the practical significance and applicability of information world-making ethics expand, emphasizing its increasing relevance in shaping human future.

Keywords AI · Information World-making · Information World-making ethics · Stipulation

T. Wang (✉)
Department of Philosophy, Center for Marxist Research in the Intelligence Age, Shanghai University, Shanghai, China
e-mail: tianen@shu.edu.cn

L. Checketts, B. S. B. Chan (eds.), *Social and Ethical Considerations of AI in East Asia and Beyond*, Philosophy of Engineering and Technology 47,
https://doi.org/10.1007/978-3-031-77857-5_8

8.1 Introduction

In contrast to the natural material world, the artificial information layer of the world is created by humans themselves. With the advancement of AI, this information layer of the world is becoming increasingly intelligent, sparking new ethical concerns regarding the making of the information layer of the world. While humans traditionally adapt to the physical environment, they now play a pivotal role in shaping the complex artificial information layer of the world, humans primarily create the environment.

Recent studies show that in big data and especially in AI, human information activities increasingly exhibit a nature of making the information layer of the world, thus the information world-making philosophy based on the reflection of it, and a new ethics, termed "information world-making ethics," has emerged, led to the increasing integration of ethics as a dimension of the information world-making philosophy, the concept of the "metaverse" embodies this excessive reflection, elevated ethical research to the level of information world-making. As a result, there is an increasingly deeper (or more basic) intrinsic connection between ethical research and religious studies, which is quite intriguing.

For humans, the information layer serves as a crucial interface, while matter remains foundational. As such, "information world-making"may be abbreviated as "world-making," with "information world-making ethics"correspondingly shortened to "world-making ethics." In this context, world-making encompasses human activities in creating information layers, while world-making ethics addresses the ethical dimensions of these endeavors, distinct from information ethics.

Information ethics deal with general ethical issues in human information activities. Whereas, world-making ethics focus on the ethical issues arising from human information world-making activities. The former mainly concerns ethical relations between individuals, while the latter is the whole human ethical problem caused by human information creation activities, involving the future and destiny of humanity as a whole.

As information civilization advances, especially AI and big data as its new foundation human activities of world-making is becoming increasingly prominent, in which the better understanding of information and its basic characteristics is getting more and more important on the one hand, and as information technology, particularly big data and AI, advances, the urgency to deepen our understanding of information grows on the other.

It is Wiener's assertion, "information is information, not matter or energy" (Wiener, 1985, 132), marks the new inception of information research. The evolution of information research has shifted from a focus on signals to sources and, ultimately, to receivers. While information research has proposed nearly two hundred definitions, they can be summarized as three primary approaches to information understanding: signal-carrying, source-preexisting, and receiver-assigning. These serve as foundational "piers" upon which a bridge towards a more accurate understanding of information can be constructed, surpassing current general information coding comprehension and reaching a receptive relation understanding. The mature form of this understanding is the receptive relation process between information receivers and sources (Wang, 2022).

Originating from the interaction between machine receivers and information sources, the unfolding of big data and AI continuously shapes the landscape of information. This not only signifies a fundamental paradigm shift, integrating philosophy and science, but also facilitates the systematic extraction of the basic characteristics of information. Systematically derived from the receptive relation understanding of information, these basic characteristics include created ex nihilo, emergence, reciprocity, and sharability. The created ex nihilo nature of information implies that human information activities constitute a form of world-making, and the human information world is entirely a result of human world-making activities. The emergence, reciprocity, and shareability of information signify that the information world created by humans possesses different characteristics from the physical world (Wang & Wang, 2024, 133–44).

By recognizing information as receptive relation, information encoding can be defined as materialization and idealization. This clarifies that information encoding is distinct from information itself, with data representing a form of information encoding. Two fundamental types of information encoding emerge: material encoding, found in physical signals and genes in DNA, and ideational encoding, seen in concepts generalizing products as natural kinds. Among them, digital coding has a special status as symbolic coding in ideational encoding of information (Wang & Wang, 2024, 101–03).

The deepening understanding of information and the clarification of its relationship with matter and energy have opened up vast development opportunities and prospects for human information activities, particularly in terms of the distinct relationships between humans and the physical and informational levels of the world.

Humans have transitioned from interpreting the world and changing the world to making the information world based on matter and energy with the development of information. If interpreting the world constitutes the "philosophy of world-explaining," then constructing the information world defines the "philosophy of world-making." The philosophy of world-making has basic characteristics that are different from the philosophy of world-explaining in terms of ways of understanding, grasping, and holistic thinking. Philosophy presents a more complete form with a symmetrical relationship between the philosophy of world-explaining and the philosophy of world-making, thereby elevating ethical research to a higher level of integrality.

In information world-making activities, "to be" and "ought to be" form an intertwined bidirectional cycle, constituting a whole. The integration between rules and laws is becoming more evident. The intertwined bidirectional cycle between "to be" and "ought to be" forms a higher-level integrated mechanism through the notion of "going to be." In the depths of intelligent algorithms, we can observe the deep mechanisms of the integration of rules and laws: facts and values are "merged" based on classified research; the development trend shows an increasing integration of rules and laws.

Traditional ethics instructs people on what can and cannot be done based on ethical principles. However, as a dimension, the ethics of AI must also guide people on how to act in a way that aligns with ethical rules and relevant laws. The law basis of ethical rules open up unprecedented the future dimension for the development of AI ethics.

In the integrated development of rules and laws, ethical governance of AI, is not merely about control but requires a deep understanding of corresponding laws and mechanisms, therefrom provides a theoretical basis and vast space for exploring the theoretical foundation of ethical support for AI development. This understanding must extend to the level of smart algorithms and broaden to enhance the understanding of the laws governing intelligent evolution on a higher level. The mutual observation of the two aspects can constitute the intelligent ethical guidance for the development of human intelligence and AI.

The integration of rules and laws means some kind of new ethics that has different properties and a comprehensive perspective. From the subjective position of AI agents to the existential risks of AI development, specific issues such as accident liability in autonomous driving cars and the existential risks posed by the development of AI must be understood and resolved more reasonably from the ethical dimension of world-making philosophy.

As information is receptive relation based on material receptivity, which is created out of nothing, rather than conserved like matter and energy, human information activities on the basis of big data have reached a level of information world-making. And when such creative endeavors involve AI, it becomes the finishing touch of "world-making." Creative works typified by AI will increasingly affect humans and giving rise to holistic ethical issues. These ethical issues not only involve a wider range of topics but also rise to the level of world-making ethics.

8.2 From "World-Adapting Ethics" to "World-Making Ethics"

From AlphaGo to ChatGPT and Sora, the subsequent evolution of AI and big data as its new foundation has fundamentally transformed the role of human activity. The fundamental transformation of human situation has correspondingly changed our ethical status. The blurring of the boundary between humans and the Creator marks the beginning of ethical issues shifting from traditional civilization to information civilization, and the speed of development since then has far surpassed our previous imaginations our previous imagination. Today, slogans like "I am crazy for creation!" abound in the media. However, theoretically, individuals must contemplate: given that humans are not Gods, how can we navigate the creation of the world? How can we create without straying from the long-term interests of humanity? What ethical consequences will there be after information world-making? The more holistic the information world-making, the sharper and more fundamental these questions become.

Information world-making on the basis of big data undoubtedly positions humans in a god-like role. Whether this world emerges through evolution or divine intervention, in the past, we could only adapt to the environment in which we lived. We attempted to understand, change, and even dominate nature, yet humans have ravaged our own habitat. Now, we are making a designed artificial information world. In the making of this artificial information world, virtual reality technology has reached a level where we can even stipulate rules that are completely different from physics, which means a series of significant ethical issues that have never occurred before. Who has the right to stipulate these rules? Who has the right to create a world with different rules from the material world? If we label the ethics necessary for adapting to this world as "world-adapting ethics," then the ethical quandaries inherent in information world creation can be termed "world-making ethics." In fact, due to the development of virtual technology, Professor Zhai Zhenming introduced the research field of "world-making ethics" several years ago and first made a clear statement:

> Because the 'physical' laws of the virtual world are artificially stipulated, this requires an academic field of 'world-making ethics,' in which we explore and formulate the 'best' set of mutually coordinated 'physical' laws in a rational way. For example, can creations in the virtual world become old? Can humans be damaged in their interactions with nature and others? Are 'natural disasters' allowed to occur in the virtual world? To answer these questions and similar ones, we rely on the birth of an unprecedented 'world-making ethics.' If we do not want to establish creations and development of the virtual world on an irrational basis that lacks rationality, we must create this academic field with a high sense of responsibility and conduct systematic and in-depth research here. (Haiqing, 2010, 163)

In the receptive relation understanding of information, virtual reality means the real information world, and the construction of virtual reality means the making of information world. Given its broader foundation, the whole range of significant ethical issues involved in big data urgently needs systematic research on the level of information world-making; due to AI's development as a finishing touch of world-making, ethics for world making is becoming more important and even urgent.

Due to the integration of human cognition and practice with the advent of the information civilization, the challenges confronting humanity are becoming increasingly comprehensive, with the ethical dimension of world-making philosophy standing out prominently. The ethical position of humanity has undergone fundamental changes. Natural world evolution does not inherently entail ethical concerns, but the artificial making of the information world not only introduces ethical issues but also renders them increasingly holistic, thereby elevating ethical inquiry to a higher level of integrality.

We are neither capable nor required to hold ourselves accountable for the creation resulting from natural evolution, even if it is attributed to a divine creator. Whether it comes from natural evolution or is created by God, we cannot discuss ethical issues for world-making. However, when humans make an information world, it is completely different; there are obviously ethical issues with world-making.

Given its close alignment with human needs, development, and progress, big data lays the foundation for creating things that satisfy human needs. However, when information world-making based on big data pose a risk to humanity itself, it runs counter to human needs (Wang, 2016). An emblematic and concerning instance is the advancement of AI surpassing the Turing point. As human information activities based on big data take human needs as the starting point and the satisfaction of human needs as the ultimate goal, big data is closely related to human needs; it is a correlation between quantities (correlated variables) and human needs, making it a multi-level correlation—a key point that distinguishes big data correlations from other correlations. This property makes big data correlations inherently related to the needs and development of humans (or human-like beings). When intelligent creations evolving based on big data pose complex ethical quandaries and even challenge human existence, the resultant ethical issues are fundamental, and the ethical risks are disruptive. This is the unique problem that the development of AI may lead to—the "humanity ethics" problem, which involves ethical issues that violate human nature, whether individually or collectively and even extend to existential risks for humanity. Behaviors in information world-making that diverge from the trajectory of human needs development represent ethical transgressions against human nature.

In an era of guiding human needs through the development of created objects, the ethics of creation activities diverge significantly from the past and even encompass the issue of shaping the evolution of human needs. The world-adapting ethics mainly maintain current social order, while the world-making ethics increasingly involve the future development of human beings. The development of AI based on big data pushes this issue to the extreme.

At the pinnacle of creation lies the creation of humans, and "creating humans" is the finishing touch of "world-making," indicating the ethical significance of AI development. "Creating humans" within the context of AI implies a transformation and extension of human relations. The development of AI will change human relations, leading to revolutionary changes in ethical relationships. The ethical implications of world-making with AI are concentrated in the ethical issues of AI creation.

8.3 The Pre-stipulation Ethics of World-Making

One of the fundamental considerations in world-making pertains to pre-stipulations. For humans, whether from natural evolution or crafted by a divine creator, the natural world already exists. However, in the case of the artificial information world created by humans themselves, the most basic issue revolves around pre-stipulation. World-making involves the pre-stipulation of design, and the human-made artificial information world is constructed on the basis of these foundational stipulations.

Under natural conditions, humans inhabit a predetermined world, with adaptation constituting the primary challenge. Grounded in big data, humans primarily inhabit the information world they've been making, which is becoming increasingly intelligent. The main challenge in this context is choice, which can even involve fundamental presuppositions and thus typically encompasses deeper ethical considerations related to world-making.

Previous studies centered on humans inhabiting the pre-existing natural world, where pre-stipulations mainly consisted of behavioral and thinking stipulations. In the process of information creation based on big data, stipulations themselves have been further extended. The activities of world-making involve the pre-stipulation of design. As a result, novel concepts such as "designing ontologies" (Little & Rogova, 2009, 70) are put forward. This is an existential stipulation that is different from behavioral and thinking stipulations, not only involving new stipulated content but also new related mechanisms.

On the one hand, world-making necessitates certain existential pre-stipulation. Without existential pre-stipulations; without these foundational precepts, world-making activities cannot proceed. On the other hand, the results of world-making also affect and even determine the direction of human evolution. From this, we can see an important new ethical issue: who determines these existential presuppositions? As Aristotle noted, choosing one possibility entails the eradication of countless others. Who possesses the authority to eliminate myriad possibilities and opt for just one in the course of human evolution? This presents an extraordinary and fundamental ethical quandary interlinked with numerous related issues.

If, in the sense of world-making, the importance of pre-stipulations, especially their impact on the direction of human evolution, means a significant ethical issue of holistic nature. How were these pre-stipulations made? What is the basis for making these stipulations? Their impacts even determine how humans develop, involving the direction of human evolution. Who can bear responsibility for making these stipulations? As human beings, we used to evolve by adapting to the environment, but now we are actively choosing the direction of our own evolution, and this choice will inevitably give rise to numerous ethical issues.

Due to the significant relationship between the pre-stipulations of artificial information world-making and human survival, shaping the future trajectory of humanity. A series of related issues will become increasingly come to the fore. Even in partial design, the influence of information tool design on humans has already been demonstrated due to path dependency. While the pre-stipulation of Windows design may not inherently possess world-making significance, Windows has, nonetheless, established a set of rules upon which subsequent behaviors are path-dependent. The irrational arrangement of keys on computer keyboards indicates that even incidental factors can shape human development.[1] Computer keyboards and railway track gauges exemplify how pre-stipulations in hardware design determine subsequent development, while the influence of pre-stipulations in software design is even

[1]The arrangement of letters on computer keyboards comes from typewriters, which in turn originated from a purely accidental phenomenon. Due to limitations in mechanical technology at the time, the keys would get tangled if typists typed too quickly. To address this, engineers came up with an irrational arrangement of letters to slow down typists. Once the keyboard became a habit, even if more rational letter arrangements were designed, they could not be widely adopted.

greater.[2] Roman numerals versus Arabic numerals even involve the selection and stipulation of cultural "memes." millions via Roman numerals demands a lifetime of learning, while Arabic numerals can be grasped by an elementary school student in a short span, underscoring the decisive influence of memes on the development of culture. It is precisely in this sense that the main focus involves the selection and configuration of predetermined choices and stipulations for world-making, which also involves the direction and design of human evolution.

The information products we design may have fundamental impacts on humanity itself. While the influence of information tool design on humans may seem subtle, the long-term accumulation of consequences may be staggering. The collapse of a constructed game world due to imbalanced order settings is not a big problem, but the artificial information world for human survival is no game. The more fundamental the pre-stipulation, the deeper its impact on the future. All of this indicates that in world-making activities, the selection and design of the direction of human evolution must have the utmost level of pre-stipulations possible. Even if a certain choice seems reasonable at present, without a holistic perspective at higher levels, it is impossible to discern clearly what the current choice means for the future. The ethical responsibility here is undoubtedly holistic, encompassing fundamental world-making ethical issues: how should pre-stipulations and corresponding rules be determined? Who determines the reasonableness of pre-stipulation? Who bears responsibility? With AI's development surpassing the Turing point, new ethical issues will arise: as the basis of norms, making basic stipulations by human beings can undoubtedly avoid human concerns about possible risks posed by AI, but the reasonableness of stipulations is increasingly a matter that requires long-term vision to determine, and it is possible to find more new ethical issues at a certain stage of development. For instance, as machine intelligence develops to a higher level and intelligent machines become more rational than humans, thus having an advantage in this field, more problems will arise: should basic stipulations be made by intelligent machines or humans? If pre-stipulations made by AI cause serious consequences, who should bear the ethical responsibility? In this sense, Plato's "philosopher-king" is not just a philosopher's "fantasy." The "philosopher-king" is more likely to be a practitioner with philosophical thinking advantages rather than a professional philosopher. Because it depends on a holistic perspective, the level of holistic perspective determines the distance of vision, so the integration of philosophy with science, society, culture, and life development will make Plato's theoretical "philosopher-king" land as a genuine "philosophical commoner."

The ethical issues stemming from pre-stipulations in world-making ethics not only involve vision but also more broadly involve complex considerations of interests based on presuppositions. These ethical quandaries stemming from such considerations often manifest in information asymmetry.

---

[2]The standard gauge of American railways determines the size of rocket boosters for American spacecraft, while the path dependence of railway gauge can be traced back to the standard gauge of electric tram tracks in the UK, the width of British carriages, and even the width of the rear ends of the two horses pulling Roman chariots.

In the era of information civilization, information symmetry stands as a foundational concern. The degree of information symmetry increasingly intersects with social fairness and justice at a basic level, while the digital gap and big data tampering are the most important issues. "The digital gap is a technological gap, that is, the achievements of advanced technology cannot be shared fairly by people, resulting in the situation of 'the rich get richer and the poor get poorer'" (Qiu, 2014). The information asymmetry stemming from the digital gap will increasingly become a basic social problem. Big data monopolies and even data tampering will artificially engender severe information asymmetry. Given its holistic nature, even small changes via smart algorithms in the relevant relationships of big data may cause significant changes, thus leading to serious ethical issues. Distorting or even tampering with big data, thus utilizing or even creating information asymmetry, is a significant ethical issue unique to the era of big data. Therefore, in the field of data research, technology and ethics are increasingly inseparable. "The data foundation of the Twenty-first century must include not only technical issues but also fairness and broader ethical issues" (Baru, 2018, 9). Big data is meant to be kept in its natural state, and distorting and tampering with it for personal gain, such as conditionally intervening in data modification, can cause very serious holistic ethical problems. And these can very easily, sometimes even only, get done by AI. Data tampering is a very serious issue because modifying data changes the correlations, which not only affect each other but are also considered to have objective significance. Injecting subjective interests into objective evidence will engenders serious ethical issues. In the making of artificial information worlds, big data monopolies and revisions will affect the whole. If the correlations of big data are changed, it may involve the holistic structure of the artificial information world; and at the information level of ecological ethics, information ecological ethics covers natural ecology and social ecology at a higher level of holistic composition, so related ethical issues will rise to the level of world-making ethics.

8.4 The Features and Characteristics of World-Making Ethics

World-adapting ethics only encompass human behavioral norms, whereas world-making ethics involve the foundation of human existence. Consequently, world-making ethics possesses distinct properties and characteristics compared to world-adapting ethics.

8.4.1 The Holistic Nature of World-Making Ethics

The holistic nature of world-making ethics stems from the broader context of world-making itself. It not only reflects the ethical implications of the holistic information world-making activities but also embodies the integration of ethical rules and world-making laws.

In cognition that mainly focuses on objects, conservation of energy and the fact that matter cannot be created out of nothing have always dominated the description of existing objects. Although there is a cycle from part to whole and from whole to part in the description, the process from part to whole is mainly highlighted, and the process from whole to part is relatively hidden. However, in the main understanding that focuses on information, due to the non-conservation and nature of information, humans have entered information world-making in a sense of creating out of nothing based on explaining and changing the world. Due to the characteristics of information world-making highlight the aspect from whole to part, or in other words, the part or the parts are created after the whole, so the most important meaning of information world-making ethics is holistic. It is precisely the ethical implications of the holistic nature of information construction that make world-making ethics have different holistic characteristics from world-adapting ethics: holistic nature involves the integrity of all aspects. As a comprehensive process of information construction, world-making means the holistic nature of world-making ethics.

This holistic nature of world-making ethics is reflected in the integration of ethical rules and world-making laws. It is the holistic nature of world-making that integrates ethical issues, thereby giving world-making ethics with a holistic nature.

The basic issues of world-making ethics are all holistic, and this holism is not only spatial and scale-based but also temporal and process-based. Correspondingly, the solution to world-making ethical problems is also based on a holistic perspective.

World-making ethics marks a higher level of holism concerning ethical issues related to AI and big data serving as its new foundation. In the AI era based on big data, all ethical issues can be attributed to world-making ethics or can be elevated to the level of world-making ethics for consideration. Whether it is emotional relationships between humans and AI machines or reconstruction of human relationships caused by this, they all have properties related to world-making ethics. From a deeper perspective, in the AI era, ethical relationships such as human relationships and simple freedom and privacy issues that are local in nature are fully integrated into the broader relationship between people and the information world they collectively create. The integration of philosophy with science, society, culture, and even everyday life in the era of information civilization makes ethical issues present different levels and forms, but overall, ethical issues aligned with information civilization era, especially AI and big data as its foundation, have characteristics related to higher-level holism, which is the most basic characteristic of world-making ethics. The holistic nature of world-making ethics imparts a fresh ethical connotation.

8.4.2 Humankind's Collective Characteristic in World-Making Ethics

The holistic nature of world-making ethics emphasizes a collective approach, encompassing humanity as a whole, including both humans and AI. Unlike the localized and individualistic nature of world-adapting ethics, world-making ethics transcends local and personal boundaries, embodying. In this regard, world-making ethics represents a form of collective ethics. Therefore, it has another basic characteristic different from world-adapting ethics: humankind's collective characteristic. The humankind's collective characteristic of world-making ethics has profound ethical consequences, the most important of which is the **reconciliation** of individual ethics with collective ethics in human society.

The ethics guiding adaptation to the world are termed "adaptive ethics," while those guiding the creation of the information world are termed "world-making ethics." The transition from "world-adapting ethics" to "world-making ethics"signifies the reconciliation of individual ethics and collective ethics, as well as the prominence of typological ethics.

The holistic nature of world-making ethics implies the reconciliation of individual ethics and collective ethics. On one hand, it signifies a shift from individual ethics to group ethics, where the interests of the group take precedence over those of individuals. On the other hand, it is not a traditional sense of individual ethics being subservient to group ethics, but rather that group ethics and individual ethics have achieved holistic integration. In other words, in the reconciliation of individual and collective ethics, group ethics increasingly take priority over individual ethics, but this priority does not conflict with individual ethics. Thus, we can see that the most basic characteristic of world-making ethics is not only its holistic nature but also its clear hierarchy, where lower-level ethics are subordinate to higher-level ethics. This is not merely about individuals yielding to the collective but rather, through hierarchical elevation, it overcomes the limitations of individuals submitting to the group and solves a series of problems brought about by individual submission to the group in world-adapting ethics. In world-making ethics, the elevation of ethical levels not only eliminates unfairness among individuals but also promotes fair treatment among individuals with different ethical needs.

8.4.3 The Commonality of World-Making Ethics

The humankind's collective characteristic of world-making ethics not only makes it correspondingly self-contained as class ethics, but also maximizes the reciprocity of information. Reciprocity of information means action symmetry, while the reciprocity of information needs means mutual benefit and harm. Therefore, on the basis of the development of AI and big data as its new foundation, the reciprocity of world-making ethics is constituted. Big data not only provides a holistic context for

AI agents but also establishes group associations and reciprocity for them. The development of AI then takes world-making to the highest level. The highest level of world-making is creating humans, and as a group, with the fusion evolution with machine intelligence, humans increasingly exist in an informational way and have stronger reciprocity (Wang, 2017). As a basic characteristic of anthropology, reciprocity embodies the basic characteristic of information. As humans exist in an informational way, their nature is ultimately reflected in information needs. The nature of information needs manifests as the commonality required by world-making ethics.

The commonality of world-making ethics not only denotes the commonality that ethical needs must meet but also means the commonality that ethical needs generate and develop. The commonality that ethical needs meet, generate, and develop implies common maintenance, common benefit and harm, where one's gain is everyone's gain and one's loss is everyone's loss. World-making ethics means that human beings choose between two opposite directions: self-disobedient of group ethics or common development of humankind's collective characteristic.

As the ethical subject is the whole of humanity, the humankind's collective characteristic of world-making ethics implies the self-contained nature of group ethics. World-making ethics is not just about human relationships, but concerns the existence of the entire existence of humankind, thus having a fundamental ontological status. World-making means that humanity sets the premise for its own existence, which not only implies completely different ethical connotations but also implies the self-contained nature of group ethics. The ethical consequences of this genetic stipulation are not just about certain individuals but concern the entire humanity. It is the self-contained nature of group ethics concerning the entire humanity that determines the self-contained nature of world-making ethics. In other words, ethical rules are not ultimately about referring to others but have a self-contained nature for the entire humanity. This humankind's collective characteristic of world-making ethics, and thus the self of group ethics, has profound ethical consequences.

In world-adapting ethics, moral behavior mainly involves some individuals in a group of people and corresponding ethical problems mainly affect some limited number of individuals. Conversely, in world-making ethics, moral behavior mainly involves the entire humankind, and corresponding ethical problems mainly affect the entire humanity. One extreme manifestation of this characteristic of world-making ethics is self-disobedient in world-making ethics. As the foundation of information civilization, big data directly links human activities and human needs and their development, making world-making not only mean the continuous creation of the human living environment but also the possibility of being surpassed by humanity's own creations, which is the most fundamental ethical problem of self-disobedient concerning the entire humanity.

The common development of humankind's collective characteristic guided by the commonality of world-making ethics can not only be linked to free development of individuals but can also herald the dawn of a new ethical paradigm. As the creation of human-like intelligent agents epitomizes the pinnacle of world-making endeavors, AI inherently involves human self-ethics at a higher level. World-making ethics has the nature of self-ethics, and self-ethics truly reveals the essence of ethics—ethics ultimately stem from self-needs, the needs of individual humans and the entire humanity. As a self-need, self-discipline in morality has the most fundamental basis. Therefore, the self-ethical nature of world-making not only provides the possibility of solving some ethical problems that are difficult to solve or even trapped in a dilemma in world-adapting ethics but also transforms the resolution of these problems into new opportunities for development.

8.5 Two Key Instances of Applying World-Making Ethics in AI

As AI advances, ethical regulations within AI applications are becoming increasingly prevalent. From addressing the existential risks posed by AI to the ethical transcendence of the responsibility gap represented by self-driving cars, two prominent levels of ethical regulation within AI emerge.

8.5.1 The Ethical Response to the Existential Risk of AI

Concerning the risks associated with AI development, the initial response naturally revolves around ethical considerations. With the development of a new generation of AI, the existential risks it brings are increasing, and the response situation has fundamentally changed. The ethical principles within AI are becoming increasingly limited as important discoveries are made. The loading of values in AI also faces significant theoretical and practical challenges, but its openness offers valuable insights into ethical responses to the existential risks of AI.

As a concept put forward at the turn of the last century, "Existential risks are those that threaten the entire future of humanity" (Bostrom, 2013, 15). These "Existential risks have a cluster of features that make ordinary risk management ineffective. The final section of this paper discusses several ethical and policy implications. A clearer understanding of the threat picture will enable us to formulate better strategies" (Bostrom, 2001). With the advancement of information technology, an increasing number of people perceive that the existential risk humanity faces primarily stems from AI, especially superintelligence. "AI would come to be among our most severe existential risks" (Armstrong, 2014, 45). Not to mention superintelligence. "Superintelligence could also be a factor in increasing existential risk" (Farquhar et al., 2017, 9). Understanding the evolution of AI and human-machine relationships at a higher level of integration of rules and patterns becomes imperative. "If the intelligence of artificial systems were to surpass that of humans significantly, this would constitute a significant risk for humanity. Time has come to consider these issues, and this consideration must include progress in AI as much as insights from the theory of AI (Müller, 2016, 1). Should AI surpass the Turing singularity and exhibit intelligence surpassing that of humans, humanity faces the existential risk of complete obsolescence. Consequently, many view superintelligence as catastrophic for humanity, underscoring the need to address AI's existential risks through the lens of world-making ethics. At a deeper level of rule and law integration, the possibility of human-machine fusion evolution and the emergence of higher intelligence levels become conceivable. Addressing AI's existential risks cannot rely solely on rules and laws or simply grasp laws; it must elevate to a higher level of rule and law integration.

On the electronic frontier, especially smart algorithm with the development of AI, on one hand, there are "coded rules", on the other hand, "code is the law. The rules governing any computer-constructed microworld—of a video game, your personal computer desktop, a word processor window, an automated teller machine, or a chat room on the network -are precisely and rigorously defined in the text of the program that constructs it on your screen" (Mitchell, 1996, 111). Which is meaningful to the research of ethics, related to that codes are the rules and codes are the laws, a further point has been made: "Code is ethics and ethics is code" (Béranger, 2018, 139). This is the concrete and vivid expression of the integration of rules and laws in the field of information technology.

This underscores that what the response to the existential risk of AI means is not a traditional ethical issue but a question related to world-making ethics, which must be based on the experiential nature of intelligent evolution and rise to the level of artificial general intelligent evolution. Dealing with the existential risks of AI cannot rely solely on rules or laws; it must ascend to a higher level of integration of rules and laws. This is not a traditional ethical issue but rather a matter related to the world-making ethics, grounded in the experiential nature of intelligent evolution and ascending to the level of generalized intelligence evolution, including human intelligence and AI.

The integrity, humankind's collective characteristic, and commonality inherent in world-making ethics, especially the integration of rules and laws it implies, open up higher-level prospects for addressing the existential risks of AI. From the perspective of world-making ethics in response to the existential risks of AI, we can see its basic principles and universal significance in the development of AI. In which humans must prioritize "Dao" over "demon": in the face of research progress with existential risks, preventive measures must be taken before implementation to ensure that there are solutions for potential risks. To ensure that the ethical reviews of making activities of AI is placed in the integrated grasp of ethical rules and the law of AI development.

8.5.2 *Bridging the "Responsibility Gap" Ethically in the Case of Self-Driving Cars*

Taking self-driving cars as a typical example, emergence of the "responsibility gap" in AI applications has become an urgent problem that needs to be solved urgently. The responsibility gap is essentially a phenomenon in which machine agents have no responsibilities at the stage when their autonomy is increasing but have not yet assumed the role of responsibility subject. With the more common application of AI, the responsibility gap brought about by AI is a theoretical subject that urgently needs to make a significant breakthrough in the field of practice. The application of AI will bring the biggest dividend that human beings can imagine at present, while the responsibility gap may constitute the bottleneck of AI application. While expecting the development of AI to bring huge dividends, it is urgent to deal with the accompanying liability dilemma. Crossing the responsibility gap in the application of AI is related to the development and application of AI, thus related to the successful delivery of unprecedented dividends in the development of AI. In the face of the responsibility gap in the application of AI, it is impossible to find a reasonable theory and method in the traditional paradigm, but to bridge it in world-making ethics which means the integration of rules and laws. In this case, related responsibilities and obligations mainly involve the design of intelligent devices, responsibility for use, and ethical obligation for group benefit. If there is no person responsible for an accident caused by a self-driving car and only beneficiary's obligation exists, relevant losses should be borne by the beneficiary group of autonomous cars in form of obligation. In terms of specific mechanisms, the immense dividends brought by the development of autonomous vehicles ultimately bear the responsibility for bridging the ethical gap, considering the outcome resolution at a higher holistic level of world-making ethics. The ethical bridge for AI application's responsibility gap will make AI application and realization of corresponding dividends form a virtuous circle.

8.6 Conclusion

The development of information technologies, particularly AI and big data as its new foundation, has not only presented ethics as an integrated dimension but has also elevated ethical research to the level of world-making. This integration has sparked deeper connections between ethical studies and religious inquiries, prompting further exploration.

The philosophy of world-making, which focuses on making artificial information worlds, introduces ethical implications that elevate ethical research to a higher level of integrality. Information is receptive relation, possessing characteristics such as created ex nihilo, emergence, and reciprocity. As human creative activities reach the level of world-making, ethical implications become increasingly holistic and profound.

The transition from "world-adapting ethics" to "world-making ethics" highlights the ethical implications of making artificial information worlds. Humans, as creators, are confronted with novel ethical issues regarding the stipulation of rules and the long-term welfare of humanity. The makings of artificial information worlds introduce unprecedented ethical issues that require systematic research and a heightened sense of responsibility.

The correlation between AI and human needs gives rise to intricate ethical dynamics, particularly as AI advances beyond the Turing point. The unique problem of "humanity ethics" arises, involving ethical issues that challenge fundamental aspects of human nature. In an era where human needs are shaped by human creations, world-making ethics extends beyond maintaining the current social order and involves directing the future development of human needs.

The integration of AI and ethics is crucial in the field of data research. The ethical implications of data tampering, information asymmetry and other ethical concerns elevate to the level of world-making ethics. Ensuring responsible and equitable practices in the making of artificial information worlds becomes imperative.

World-making ethics embodies holistic nature, humankind's collective characteristic, and commonality. Holistic nature arises from the holistic nature of world-making itself, integrating ethical rules and world-making laws. Humankind's collective characteristic prioritizes group ethics over individual ethics, reconciling individual and collective ethics while emphasizing the hierarchy of ethics. The commonality of world-making ethics generates reciprocal relationships among AI agents, emphasizing shared maintenance, benefits, and harms.

The application of world-making ethics in AI is evident in various cases, such as addressing existential risks and bridging the "responsibility gap" in AI applications. By incorporating world-making ethics, preventive measures can be prioritized to address existential risks. Bridging the responsibility gap requires an ethical approach that integrates rules and laws, ensuring that responsibilities and obligations related to the design and use of intelligent devices are upheld.

The integration of ethics into the realm of world-making in AI engenders profound implications for ethical research and decision-making. It calls for a comprehensive perspective in addressing ethical issues, considering the holistic nature of artificial information world construction. By upholding world-making ethics, ethical and responsible practices can guide the development and application of AI.

The development of AI has brought about significant changes in human activities and ethical considerations. The philosophy of world-making, which focuses on making the information worlds, introduces ethical implications that elevate ethical research to a higher level of integrality. As humans take on the role of creators, new ethical questions emerge regarding rule-setting and humanity's long-term interests. Unlike natural evolution, artificial information world not only involves ethical considerations but also makes them increasingly holistic. The emergence of "humanity ethics" poses ethical challenges that touch upon the essence of human nature. The transition from world-adapting ethics to world-making ethics highlights the integration of ethics as a dimension of information world-making philosophy. This integration has deeper intrinsic connections with religious studies, expanding the scope of ethical research. The philosophy of world-making emphasizes humankind's collective characteristic and commonality of ethics besides integrity. The application of AI in addressing existential risks and bridging the "responsibility gap" exemplifies the practical implementation of world-making ethics. These cases require a holistic perspective and the integration of rules and laws to address ethical challenges and ensure responsible and beneficial development of AI. Ultimately, the development of AI and making of artificial information worlds necessitate integration of technology and ethics, fostering responsible and equitable practices for the betterment of humanity.

Funding The National Social Science Fund of China (Major projects): Research on Digital Intelligence Technology with Philosophical Development and Knowledge Production Paradigm Change.

References

Armstrong, S. (2014). *Smarter than us: The rise of machine intelligence*. Machine Intelligence Research Institute.

Baru, C. (2018). *Data in the 21st century. Vladik Kreinovich, Songsak Sriboonchitta, Nopasit Chakpitak ad., predictive econometrics and big data*. Springer International Publishing AG.

Béranger, J. (2018). *The algorithmic code of ethics: Ethics at the bedside of the digital revolution*. ISTE Ltd and John Wiley & Sons, Inc.

Bostrom, N. (2001). Existential risks: Analyzing human extinction scenarios and related hazards. *Journal of Evolution and Technology, 9*. Available at http://jetpress.org/volume9/risks.html

Bostrom, N. (2013). Existential risk prevention as global priority. *Global Policy, 4*(1), 15–31.

Farquhar, S., et al. (2017). *Existential risk: Diplomacy and governance, global priorities project*. Future of Humanity Institute. University of Oxford.

Haiqing, J. (2010). Ethical issues regarding the expansion of virtual worlds–an interview with professor Zhenming Zhai. *Philosophy Analysis, 3*, 160–165.

Little, E. G., & Rogova, G. L. (2009). Designing ontologies for higher level fusion [J]. *Information Fusion, 10*, 70–82.

Mitchell, W. J. (1996). *City of bits: Space, place, and the infobahn*. MIT Press.

Müller, V. C. (2016). Editorial: Risks of artificial intelligence. In *Risks of artificial intelligence*. CRC Press.

Qiu, R. (2014). Ethical issues of big data technology. *Science and Society, 1*, 36–48.

Wang, T. (2017). The existential modes of information for individuals in the information civilization and its philosophical implications. *Philosophy Analysis, 4*, 18–30.

Wang, T., & Wang, X. (2024). *Information as receptive relation*. Routledge.

Wang, T. (2016). Causality in big data and its philosophical implications. *Social Sciences in China*, 6, 22–42.

Wang, T. (2022). Contemporary manifestation of information and its fundamental characterastics, *Social Sciences in China*, 1, 90–113.

Wiener, N. (1985). *Cybernetics, or control and communication in the animal and the machine* (2nd ed.). MIT Press.

Chapter 9
Artificial Intelligence Liberalism: Questions Small and Big

Carl Mitcham

Abstract Current ethical discourse on AI ethics is too narrow. The argument divides into three parts. Part one offers a simplified overview that samples recent work by Mark Coeckelbergh, one of the few philosophers who properly seeks to combine ethics and politics. Part two pushes for granting existential risk a more prominent, limit situation role in AI ethical-political discourse, and call attention to the post-World War II emergence of concerns about prospective potentials for global technogenic catastrophies. Part three pays tribute to Toby Ord's effort to construct the kind of bridge that I advocate, but then questions whether the liberalist political philosophical basis on which it rests. Part four calls out liberalism as enabling AI in ways that undermine efforts to subject it to proper ethical-political judgment and regulation.

Keywords AI · Ethics · Existential risk · Political philosophy · Politics · Liberalism

9.1 Introduction

My thesis is that ethical reflection on AI, as currently practiced, is often too limited in scope. It focuses mostly on small or well-bounded questions. We need to ask bigger questions—questions that take us from ethics to politics, and confront us with what Karl Jaspers termed "limit situations."

AI research, development, and marketing are one gigantic gamble—a gamble interwoven in with multiple other gambles—to create a tapestry of strengths, weaknesses, opportunities, and threats that exceed those being subjected to piecemeal ethical interrogation. The temptations and risks are global, historical, and perhaps

C. Mitcham (✉)
Colorado School of Mines, Golden, CO, USA
e-mail: cmitcham@mines.edu

L. Checketts, B. S. B. Chan (eds.), *Social and Ethical Considerations of AI in East Asia and Beyond*, Philosophy of Engineering and Technology 47,
https://doi.org/10.1007/978-3-031-77857-5_9

existential. It is imperative to expand ethical and political discourse on what we call artificial intelligence beyond the customary parameters.

I begin with a simplified review of existing AI ethical and political philosophical discourse in order to sketch the intellectual landscape. I then make a case for taking more seriously than is customary (at least in sober academic discourse—popular and public intellectual discourse can be more alarmist) potentials for existential catastrophe: for considering the implications of possible disruptive upheavals in the landscape. Finally, I speculate on why it is so hard in our time to think seriously about big questions of temptation and threat. The controlling element in the narrative, I suggest, is liberal democratic theory, which manifests locally as what I propose to call artificial intelligence liberalism.

9.2 Orientation: Customary Parameters

But first the existing or customary parameters: For this I'll make use of Mark Coeckelbergh's companion volumes *AI Ethics* (2020) and *The Political Philosophy of AI* (2022a). Together they exhibit a broad command of the literature and provide good if glancing accounts of the current state of discourse on the ethical-political in AI research, development, and application. Additionally, both volumes argue that ethics alone is not enough (see also Mitcham, 2015, 2020). All ethical questions entail political ones, just as Aristotle's *Nicomachean Ethics* and *Politics* constitute one sustained philosophical inquiry.

The first volume offers a philosophical journalistic account of apocalyptic fears and transhumanist dreams, followed with a more sober account of debates about human-machine differences and the possibility of AI programed morality. A non-technical account of machine learning and data science leads into questions of privacy, responsibility, delegated decision making, transparency, bias, and the future of work in an AI economy. Without going through contortions of aping social science research methods (as in Stahl & Eke, 2024), Coeckelbergh provides a literate account with a few modest policy proposals for addressing problems, arguing for ethically sensitive design practices that incorporate democratic values and visions of the good life.

The second volume goes further. Setting aside epistemological issues of what AI knows and how it knows it—as well as metaphysical questions about what AI really is—Coeckelbergh identifies five themes for AI ethical political philosophy:

- Freedom,
- Equality and justice,
- Democracy,
- Power, and
- Anthropology.

In what ways does AI enhance or not individual freedom, equality, justice, and democracy? What kind of personal, corporate, and state power does AI exercise or enable? What does the human ability to design and construct AI systems tell us about human nature, if there is such a thing? How does AI challenge our conceptions of what makes us human? What might it tell us about non-humans and what we once called the natural environment?

The pivotal issue for Coeckelbergh, rightly, is power. His power chapter is the longest one in the book, which he remarks was initially titled "Artificial Power." Indeed, we might ask whether the exercise of power, by any means other than reason, rhetoric, or social psychological pressure—that is, with anything from a stick, to a gun, to an atomic bomb, to an algorithm—could not be called "artificial power." Insofar as the most emphatic form of power is violent force, we are reminded that the military is a prominent driver of AI research and development. Warfare is politics by other, more emphatic means.

To develop the power argument, Coeckelbergh draws on conceptual distinctions limned in Faridun Sattarov's *Power and Technology* (2019). Sattarov identifies four categories of technological power: episodic, dispositional, systemic, and constitutive. AI power manifests in all four categories.

Episodic power is exercised when AI seduces, coerces, or manipulates in a one-off, via social media and recommender systems—or in AI enabled acts of violence and destruction.

Dispositional AI power increases an individual's power over other people and the natural environment, especially when it is aligned with some human interest. Generative AI can substitute for human labor both physical and mental.

Systemic AI power supports neoliberal capitalism, authoritarianism, and other economic-political orders, most obviously if unobtrusively in the form of surveillance technologies. Platform capitalism under supervision of algorithmic productivity programs create whole new configurations of oligarchic power. Financial engineering using AI guides capital flows across sectors and national borders.

Questions of systemic power bleed into what is called the "AI alignment problem"—an issue somewhat slighted by Coeckelbergh, but one I return to later.

Finally, according to Coeckelbergh (2020), "AI can play a role in the constitution of the self and the formation of the subject, even if we are not aware of it" (100). Constitutive power creates, forms, and transforms social actors themselves—from within, as it were. Most of his power analysis interrogates this fourth type, challenging the adequacy of Marxist criticisms of technocapitalism. Instead, Coeckelbergh draws on Michael Foucault's notions of biopower and technologies of the self, complemented with Judith Butler's concept of the performative self, to examine how AI mirrors and generates new liquid identities—and what might be called mass narcissistic expressionism. Central to his concern are the ways surveillance and self-disciplining through the algorithmic management of big data (by self or others) constitute quantum leap transformations in liberal democratic individualist autonomy.

Coeckelbergh concludes his questioning of AI and power by claiming that "talking about power and AI enables us to bring in social and political theory that helps to conceptualize the political aspects of AI" (122–123). I want to turn this around, and to argue that the pressure of political issues generated by AI—those related to freedom, justice, democracy, and power—are greater than what existing social-political theory can adequately address. In concert—and especially if we add an issue existential temptation and threat—AI pushes critical thinking toward questioning the contemporary political order—at least as found in Europe and North America.

Coeckelbergh's last chapter on the possible implications of AI for philosophical anthropology picks up a discussion from the *AI Ethics* book. He suggests, for instance, that robots might at some point become so sophisticated that they need to be included within the political realm by being granted rights. But the more important debates about transhumanism—whether the symbiosis of mind and computer is producing "Humanity 2.0," as claimed by Steve Fuller (2011), or the self-destruction of humanity, as argued by Robert Frodeman (2019)—remain on the fringe. (It should nevertheless be noted that Coeckelbergh is a prolific author in the AI ethical-political arena. The two books referenced here are complemented by four others related to AI: See Coeckelbergh, 2022b, c, 2023, and 2024.)

9.3 Taking Existential Risk Seriously

Obscured by a focus on the plethora of particular social and political issues is a really "big" one: the combination of temptation, threat, and catastrophic existential risk. AI is what J. Robert Oppenheimer called a "sweet" or tempting technology, hard to resist. It is likewise a two-faced threat: If you don't adopt it, someone else will; but adoption can run the risk of killing you." When I characterize as "small" the set of social issues that constitute the customary parameters, I don't mean they're not serious. They are serious, and it's imperative to address them. But they must needs be addressed in light of the bigger one of existential risk—especially in the ways they interact and are linked together with others such as nuclear warfare, the continuous flooding of new chemicals and wastes into the environment, biodiversity loss, and global climate mutation, synthetic biology, rather than simply in the cautious, piecemeal way that tends to predominate. Only in passing does the existential risk that AI poses for human life as we know it get attention—and even then mostly to be downplayed or rejected as fear mongering.

As background for thinking about the big issue of catastrophic existential risk—an issue in which AI plays its own special role—let me digress into a selective run through history on the emergence of existential risk discourse that is associated with science, engineering, and technology—a discourse in which AI has come to occupy an increasingly prominent place.

Thinking about existential risk originated with recognition of the potential catastrophic consequences of nuclear warfare. Immediately after the bombing of Hiroshima atomic scientists and engineers organized to educate politicians and the public about the dangers of nuclear weapons. One key effort was founding of *The Bulletin of Atomic Scientists,* which in 1947 began featuring on its cover a

"Doomsday Clock" to graphically represent the danger. The closer to midnight on the clock, the closer to existential catastrophe.

Originally set in 1946 at 7 minutes to midnight, in 1953 with explosion of the first hydrogen bomb it advanced to 2 minutes to midnight. At the end of the Cold War in 1991 it dropped back to 17 minutes from midnight. Since 2010 it has been advancing again until as of January 24th 2023, reflecting the War in Ukraine, the clock now stands only 90 seconds from midnight, the closest it has ever been.

Pushing back against what he considered such excessive fear mongering, in 1960 Herman Kahn's *On Thermonuclear War* argued for more analytic "thinking about the unthinkable" (his term). According to Kahn, nuclear war need not be an existential threat, at least for the United States. It could be survivable, if only we would engineer for survival by creating bomb shelters and other structures. Kahn and associates argued for using game algorithms, in anticipation of artificial intelligence, to manage nuclear deterrence.

The same year conservation biologist Rachel Carson in *Silent Spring* (1962) shifted concern from nuclear weapons to chemical pollution, from physics to biology. In the process she helped jump start the contemporary environmental movement. Creation of the U.S. Environmental Protection Agency in 1970 was one signat achievement. Since Carson, however, conservation biology has become increasingly entangled with molecular biology and, through that, with the use of various forms of computer simulation and artificial intelligence that proposes "to save the phenomena" by means of algorithmic discovery and redesign—running a new kind of risk, that of turning nature into something we only see on the screen or displace with what Christopher Preston (2018) celebrates as "our synthetic age."

Then around the turn of the millennium, as the world became aware of the potential catastrophic consequences of stratospheric of ozone depletion and global climate change, Canadian philosopher John Leslie, in *The End of the World* (1996), parsed a spectrum of recognized global catastrophic risks. The basic distinction was between natural disasters such as asteroid impacts and human caused disasters such as nuclear war. But more important than such recognized risks, Leslie argued, were unrecognized ones. These included dangers as diverse as genetically engineered organisms and the appointment of an apocalyptist Christian as Secretary of Defense, someone who might believe that nuclear apocalypse was the will of God.

The next decade witnessed a rapid growth in discussion about existential catastrophic risk, no doubt dragged along by the 2000 proposal to conceive the world in terms of a new geologic era called the Anthropocene, that is, as a scientifically observed and humanly engineered geological era. One proposal for a planetary date stamp for the beginning of this new geological era is the layer of radioactivity laid down around the globe by nuclear weapons testing from 1945 to the mid-1960s.

In England research on existential catastrophic risk was popularized when, in 2003, British Royal Astronomer Martin Rees published *Our Final Century: Will the Human Race Survive the Twenty-first Century?* Rees argued that human survival is threatened by the unintended destructive effects of technology (e.g., nuclear weapons, chemicals pollution, nanotechnology, bio- and genetic engineering, artificial intelligence) and potential terrorist or malign uses.

Two years later Swedish philosopher Nick Bostrom founded the Future of Humanity Institute (FHI) at Oxford University. (FHI eventually fell out of favor with Oxford and was closed in 2024.) In Bostrom's analysis, existential risk is defined as "one in which an adverse outcome would either annihilate Earth-originating intelligence life or permanently and drastically curtail its potential." In 2012 Martin Rees founded a complementary (and competitive) Centre for the Study of Existential Risk at Cambridge University, and in 2018 updated his book as *On the Future: Prospects for Humanity*.

Meanwhile in 2014 in the United States MIT cosmologist Max Tegmark and Skype co-founder Jaan Tallinn, with endorsements from such luminaries such as Stephen Hawking, Elon Musk, Alan Alda, Morgan Freeman, and others, created The Future of Life Institute (FLI) as an NGO of educational outreach and lobbying to reduce global catastrophic and existential risks. The FLI emphasizes existential risk from advanced artificial intelligence but includes as well as risks from biotechnology, nuclear weapons, and climate change. FLI was the first institute to highlight AI as the leading-edge enhancing risks across multiple risk areas. Shortly after founding FLI began issuing a series of "Open Letter" public policy advisories: on AI research priorities (October 2015), autonomous weapons (February 2016), Asilomar Principles (August 2017), lethal autonomous weapons (June 2018), calling for a pause in giant AI experiments (March 2023), and AI licensing (October 2023). It may be noted, however, that as of mid-2024 the much publicized letter calling on AI developers to slow things down in order to institute a framework for ensuring safety or, failing that, for governments to step in with a moratorium, has not has not received much traction.

9.4 The Precipice

The single most comprehensive product of the Oxford Institute research program is Australian philosopher Toby Ord's *The Precipice: Existential Risk and the Future of Humanity* (2020). The book has its own website and has been widely and favorably reviewed, in places as diverse as *Science* magazine, *Notre Dame Philosophical Reviews*, and *The New Yorker*. Ord's thesis is that humanity currently exists on a dangerous precipice of potential civilizational collapse or global existential catastrophe.

Ord begins by distinguishing between local and global existential catastrophe. In each case civilizational collapse can be partial or total. Partial collapse would leave some humans alive but reduced to a subsistence level. Full collapse would leave no human presence. As examples at the local level, compare what happened to the Easter Island population after they destroyed their environment and the Anazazi at Mesa Verde or the Chernobyl Exclusion Zone after the 1986 nuclear disaster. Ord argues for what he calls "longtermism." Longtermism "takes seriously the fact that our own generation is but one page in a much longer story, and that our most important role may be how we shape—or fail to shape—that story." In his view,

> One doesn't have to approach existential risk from this direction — there is already a strong moral case just from the immediate effects — but a longtermist ethic is ... especially well suited to grappling with existential risk. For longtermism is animated by a moral re-orientation towards the vast future that existential risks threaten to foreclose. (46)

From the longterm perspective, there are three great threat categories, similar to Leslie's: currently existing natural risks, currently existing anthropogenic risks, and risks of future technological developments. Existing natural existential risks include asteroids and comets, super-volcanic eruptions, and stellar explosions. These are in some sense the most existential but the least able to engineer our way out of, and Ord spends the least amount of time on them—although he does mention NASA's work detecting asteroid threats and designing deflection mechanisms.

Existing anthropogenic threats include nuclear weapons, climate change, and environmental damage. Writing in 2020 Ord dismissed any return to a nuclear cold war as unlikely, a view that was and is perhaps too sanguine. The Russian threat to use them in Ukraine, and the existence of multiple nuclear weapons states hostile to the liberal world order, with others on the cusp of acquisition, has in just 4 years called Ord's assessment into question.

His discussion of climate change relies heavily on the 2014 IPCC report and what is now generally recognized as the false optimism of the 2015 Conference of Parties (COP) Paris Agreement. Not only was the Paris Agreement judged wholly inadequate by *Science* magazine, which ran the numbers and said they didn't add up. Even that inadequate accord is not being lived up to by the signatories. The COP process was, to be blunt, a fraud and produced a lot of well-intentioned liberal fake news.

The discussion of environmental contamination has also been quickly outdated. For example, it contains no mention of persistent organic pollutants (POPs) or "forever chemicals." Ord nevertheless concludes that

> Nuclear war, climate change and environmental damage are extremely serious global issues — even before we come to the question of whether they could cause existential catastrophe. In each case, humanity holds tremendous power to change the face of the Earth in ways that are without precedent in the 200,000 years of *Homo sapiens*. (119)

In accord with his own longterm perspective, it is to future threats that Ord devotes the most attention. Future threats include pandemics, unaligned artificial intelligence, and various dystopian scenarios. Once again, the experience of the covid-19 pandemic challenges some sanguine assumptions of Ord's analysis, especially the idea that technical solutions will be rationally and readily adopted.

Ord doesn't leave things at the conceptual level but surveys a large number of empirical studies related to each category. In his 421-page book, 170 pages (or 40%) is composed to detailed technical notes and references. Ord expends considerable analytic energy trying to develop quantitative risk assessments for different scenarios. By his estimate, "we face about a thousand times more anthropogenic risk over the next century than natural risk" (87). Indeed, his estimate is that risks from nuclear war, climate change, and environmental damage are each orders-of-magnitude higher than all natural risks combined.

Integrating the risk landscape, Ord estimates a 1 in 6 total risk of existential catastrophe in the next 100 years. To drive home his argument here, he quotes Winston Churchill—arguably the most consequential British Prime Minister of the last two or three centuries—from a speech he made in 1946, post atomic bomb, at Westminster College in Fulton, Missouri, and broadcast on American radio.

> The Dark Ages may return, the Stone Age may return on the gleaming wings of Science [and technology], and what might now shower immeasurable material blessings upon mankind, may even bring about its total destruction.

For Ord any risks associated with artificial intelligence research, development, and application are not properly separated from risks involved with other increasingly AI-enhanced scientific and technological developments: from nuclear science and engineering, chemical science and engineering, geoscience and engineering, bioscience and engineering. AI is now woven into and transforming all projects and enterprises—turbo charging them with speed and money, temptation, and threat. Recent developments in generative AI once again up the ante.

As emphasized at the beginning, AI research, development, and marketing are one gigantic gamble—a gamble interwoven in with multiple other gambles—a collaborative gamble whose risks greatly exceed those being subjected to piecemeal ethical interrogation.

In the face of this gamble with the future, Ord introduces a note of salutary caution by means of personal reflection. As he writes,

> Growing up, I had always been strongly pro-technology. If not for the plausibility of … catastrophic risks, I'd remain so. But instead, I am compelled towards a much more ambivalent view. I don't for a moment think we should cease technological progress…. We should [act] carefully, and use some significant fraction of the gains from technology to address potential dangers, ensuring the balance stays positive. (123)

Ord even goes so far as to propose a monetary value for the fraction of science and engineering wealth production that we should invest in researching potential dangers so as to exercise better governance: at least as much as we spend each year on ice cream (63).

Ord reiterates the issue in his penultimate chapter, "Safeguarding Humanity," with a version of the cultural lag thesis. The concern that the expansion of technical prowess is taking place faster than our will or wisdom to manage it has become a standard trope since the Romantics. As Winston Churchill, again, had observed in 1930, "while men are gathering knowledge and power with ever-increasing and measureless speed, their virtues and their wisdom have not shown any notable improvement as the centuries have rolled" (Churchill, 1932, 202).

Echoing Churchill, Ord writes that "the current predicament stems from the rapid growth of humanity's power outstripping the slow and unsteady growth of our wisdom" so that

> a more patient and prudent humanity would … try to limit this divergence. Most importantly, it would try to increase its wisdom. But if there were limits to how quickly it could do so, it would also make sense to slow the rate of increase in its power — not necessarily putting its foot on the brake, but at least pressing more lightly on the accelerator. (206)

But now a question arises: Given the evident risk of a continued pace of scientific research and engineering development—especially and prominently in artificial intelligence—why is this not happening? Why no moderation? Asking this question opens the door to political philosophy and the fourth, short, final part of my argument.

9.5 The Problem of Liberalism

The first part of my talk offered a simplified overview of AI ethics and politics that modestly prioritized politics. The second part argued for a strong prioritizing of risk, especially global existential risk, and the linking of AI into a manifold of other scientific and engineering activities. Let me now consider why it might be that recognition of global existential risk has had and continues to have little to no impact in slowing down or moderating the AI and associated research and development juggernaut even when faced by repeated public appeals by leading AI intellectuals and extended arguments by such figures as Deep Mind co-founbder Mustafa Suleyman (2023).

Ord's policy proposal for slowing the momentum of scientific research, development, and application is not something easily enacted. Science and engineering may be able to reduce the velocity at which a vehicle accelerates or limit its speed by means of a feedback loop cybernetic governor. Science and engineering are able to design processes and products that moderate whatever pollution or waste they generate by enhancing efficiency. Science and engineering can create whole new disciplines of risk analysis and safety engineering that aim to reduce fragilities and dangers in the products and systems of our engineered and engineering world. What they cannot of themselves do is persuade people or marshal political action to reduce or moderate what can only be described as a social-political addiction to discovery and innovation. We simply cannot stop giving way to temptation; fear gets little traction even while it shadows all actions. Philosophical arguments as diverse as those of Hans Jonas (1984) and Jean-Pierre Dupy (2022) seem to go nowhere. The parameters influencing adoption are already set in place by a cultural and political commitment to science and engineering. Adoption is an issue ultimately of politics—most often, in our world, mediated by economics.

As soon as Ord broaches the engineering power vs wisdom gap, however, he sets it aside. He simply calls for more economic investment in research, effectively admitting the impotence of his argument with regard to the most important issue. This mirrors his stance toward natural existential risks such as asteroids and stellar explosions. Although these are the most existential threats, as it were, they are the least able to technoscience engineer away, so Ord didn't spend much time on them. His emphasis defaulted to anthropogenic or human generated risks. The political question of how to reduce the on-going momentum in anthropogenic existential risk production—in flattening or widening what he calls "the precipice" on which we are perched—is in a similar manner slighted, in favor of more technoscientific and engineering research. There remains a persistent, however qualified belief that engineering and technology can save us from the expanding spectrum of problems and threats, whether existential or not. Yet on the basis of Ord's own analysis and admission, this would seem to be an unwarranted faith.

The root political theory supporting and animating AI research, development, and application is, I submit—returning to Coeckelbergh's claim about the need to draw on political theory—liberalism: liberalism being that political theory that priorities the freedom and equality of individuals. Political philosophical criticism of AI demands questioning of this ideology because the AI community is locked in a symbiotic embrace with liberalism, the political theory that takes the freedom of equal individuals as the primary reality.

The argument for atomistic individualism (in the first instance, by philosopher Thomas Hobbes) became the received starting point for all classic modern political theories (from John Locke and Jeremy Bentham to Jean-Jacques Rousseau). In liberalism the fundamental good, the *summum bonum*, is freedom (liberty). All other goods are created by individuals deciding what they want. The fundamental challenge of liberalism is to construct a social organization among individuals that maintains freedom and becomes productively supporting of any other humanly desired goods.

The social organization of the scientific research and development community meets this challenge more completely than any other. The scientific enterprise assumes individual scientists ideally maintaining their freedom to research and develop in any way they choose, while freely submitting to the assessment of their work by other free individuals on the basis of impersonal criteria (especially reproducibility) in a manner that, as a matter of historical fact, has been incredibly productive of both knowledge and (eventually, through capitalist engineering uptake) material culture consumer goods. Although the liberal political order does not always live up to its ideals—cheaters we will have with us always—it is remarkably resilient.

So on the one side, science models liberalism. But on the other, it depends for the toleration and even the support of its existence on a liberal state. Science unintentionally works like an acid to eat away at many otherwise common, stabilizing social beliefs, opinions, and traditions. The most obvious example is religion. Science undercuts many religious beliefs. But the churn of evidence-based health and medical knowledge production also sponsors disagreements between older and younger generations, and scientists self-assert an inherently international community that devalues national patriotisms. For the scientific community to thrive, its implicit and sometimes quite explicit prioritizing of international over national relationships—that is, scientific cosmopolitanism—must be politically suffered in silence, in the name of a freedom that can easily impress non-scientists as elitist and socially destructive.

More than this, the elitist scientific community must be funded by the larger, surrounding non-scientific society. Isaiah Berlin famously distinguished two types of liberty: freedom-from (or negative liberty) and freedom-for (positive liberty). It's not enough for scientists to be tolerated, to have freedom-from interference by society. The larger society must fund science. Without the social and political funding of science from without by the larger non-scientific world, scientists would inhabit a poor, much less productive community. The larger world must create freedom-for or the freedom-to practice science by paying scientists for their work. Fortunately, it turns out that science can in many instances be monetized by engineering capitalism. And even more fortunately, in a social order where all individuals are free to seek the satisfaction of any and all of their own personal desires, their competition to do so by any means available creates a demand for the ever more aggressive monetizing and commercialization of science.

To close my argumentative circle, the problem here might earily be described as a social analogate of the so-called AI alignment problem (for which the best straight forward introductory analysis may be the *Wikipedia* article: Anonymous, 2024; it is certainly less long-winded than Christian, 2020). The AI alignment warning problem was first clearly stated in 1960 by Norbert Wiener: "If we use, to achieve our purposes, a mechanical agency with whose operation we cannot interfere effectively… we had better be quite sure that the purpose put into the machine is the purpose which we really desire" (Wiener, 1960). With AI, the apparent liberty that it may have in operationally processing its own goals, whether these might be programmed into it or developed internally by means of program evolution, creates at least two ways for it to become detached from and potentially even opposed to human benefit: goal literalism and goal drift. But these same two mechanisms can lead for the scientific community and its research processes, insofar as it is structured about liberalist free self-determination ideals, can become misaligned from social benefit to the liberal society in which it is politically embedded.

9.6 Conclusion

So what's the conclusion?

I've argued that AI ethical and political discourse has been too narrow or small-minded, that we need to recognize AI as contributing to global existential risks, especially through its synergistic interactions with the manifold of techno-engineered and engineering mutations of our engineered and engineering world. Then I proposed that we are failing to act on such a big picture, big problem recognition because of the resistance of liberalism to countenance any restrictions on freedom, especially freedom in the practice of AI research, development, and application. So what is to be done?

Quite frankly, I'm not sure there is much that can be done—there are only things that can be thought. There are no real prospects for a Leninist response to the question, that is, the organization of an advance guard to lead a revolution to topple the system. Which is not to say that we shouldn't try to chip away around the edges as best we can, be willing to commit to the revolution if a movement should form, while making ready to face the looming challenges that wait in the wings—and not solely in the wings—of artificial intelligence.

Remember the old blockbuster movie serial, "The Perils of Pauline," about a damsel repeatedly in distress, quintessentially tied to the railroad tracks by an evil kidnapper? Pauline is a cartoon figure for a person in the grip of what Karl Jaspers termed a "limit situation" that has the potential to disrupt inauthenticity and facilitate self-knowledge. Without intending to be flip—we should always try to realize that there is something comical as well as tragic about the human condition, even about AI—I've sometimes thought of us as tied down on a our own track with a train approaching in the distance, and us considering our options. We can struggle with the chains that confine us, hoping against hope that we will be able to free ourselves in time. In the face of a looming catastrophe there is the option of more research and development, struggling to figure a way out, what I've also called the MacGyver option, referencing a 1990s American TV serial about an engineering genius hero secret agent who always at the last-minute invents his way out of an impossible fix.

Then there is the option of simply looking the other way. Don't think about that train approach from the east, look down the open tracks to the open west, and just think positively about the future. This is the power of positive thinking, which might also be called the head-in-the-sand option.

Against the MacGyver and ostrich options I propose a third, philosophical one: looking as open-eyed and clear-mindedly as possible at the on-coming train, practicing a kind of contemplative appreciation of our fraught state and the inescapable limited character of the human condition. This would be an austere and uncommon exercise with only the most exacting consolations of knowing for its own sake as a good-in-itself, a raising of one's consciousness to a level that perhaps could, in that very raising, become a kind of transcendence of the catastrophe—a catastrophe that we in fact have called down upon ourselves. After all, we tired ourselves to the track, in the belief that a train would never come; that this track didn't have trains on it anymore. It would necessarily include a recognition, appreciation, and contemplation of the ways in which it was not fate that tied us to the tracks. We would thus, with sobering probity, become who we are.

Whether and to what degree it is possible or even desirable to entertain such an option is, I would suggest, the genuinely big philosophical ethical-political question of artificial intelligence.

N.B. Some sections of this text undoubtedly draw heavily on related iterations of my argument.

References

Anonymous. (2024). "AI Alignment," *Wikipedia*, accessed May 31, 2024.
Carson, R. (1962). *Silent Spring*. Houghton Mifflin.
Christian, B. (2020). *The alignment problem: Machine learning and human values*. Norton.
Churchill, W. (1932). *Thoughts and adventures*. Thornton Butterworth Limited.
Coeckelbergh, M. (2020). *AI ethics*. MIT Press.
Coeckelbergh, M. (2022a). *The political philosophy of AI: An introduction*. Polity Press.
Coeckelbergh, M. (2022b). *Robot ethics*. MIT Press.

Coeckelbergh, M. (2022c). *Self-improvement: Technologies of the soul in the age of artificial intelligence*. Columbia University Press.
Coeckelbergh, M. (2023). *Digital technologies, temporality, and the politics of co-existence*. Palgrave Macmillan. ISBN 978-3-031-17981-5. OCLC 1342984554.
Coeckelbergh, M. (2024). *Why AI undermines democracy and what to do about it*. John Wiley.
Dupy, J.-P. (2022). *How to think about catastrophe: Toward a theory of enlightened doomsaying*. Michigan State University Press.
Frodeman, R. (2019). *Transhumanism, nature, and the ends of science*. Routledge.
Fuller, S. (2011). *Humanity 2.0: What it means to be human past, present and future*. Palgrave Macmillan.
Jonas, H. (1984). The imperative of responsibility. In *Search of ethics for the technological age*. University of Chicago Press.
Kahn, H. (1960). *On thermonuclear war*. Princeton University Press.
Leslie, J. (1996). *The end of the world*. Routledge.
Mitcham, C. (2015). Ethics is not enough: From professionalism to the political philosophy of engineering. In S. S. Sethy (Ed.), *Contemporary ethical issues in engineering* (pp. 48–80). IGI Global.
Mitcham, C. (2020). The ethics of technology: From thinking big to small—and big again. *Axiomathes, 30*, 589–596.
Ord, T. (2020). *The Precipice: Existential risk and the future of humanity*. Bloomsbury.
Preston, C. J. (2018). *The Synthetic Age: Outdesigning evolution, resurrecting species, and reengingeering our world*. MIT Press.
Rees, M. *Our final hour: A scientist's warning: How terror, error, and environmental disaster threaten humankind's future in this century—On earth and beyond*. Basic Books.
Sattarov, F. (2019). *Power and technology: A philosophical and ethical analysis*. Rowman and Littlefield.
Stahl, B. C., & Eke, D. (2024). The ethics of ChaptGPT – Exploring the ethical issues of an emerging technology. *International Journal of Information Management, 74*, 102700.
Suleyman, M. (2023). *The Coming Wave: Technology, Power, and the Twenty-first Century's Greatest Dilemma*. With Michael Bhaskar. New York: Crown.
Wiener, N. (1960, May 6). Some moral and technical consequences of automation. *Science, 131*(3410), 1355–1358.

Chapter 10
Resetting Machine Ethics: Rationalism, Hypocrisy, Disagreement, and the Skillful-Expert Model

Felix S. H. Yeung and Fei Song

Abstract Existing approaches to machine ethics harbor an unquestioned commitment to the development of ethical machines and an unreflective optimism that ethical principles can be executable by machines. The first part of this paper raises two challenges to such dogmas: the hypocrisy challenge and the disagreement challenge. The first challenge is that, aside from finding the right machine ethics program, machine ethicists must consider whether their development of such machines is consistent with the precepts of their adopted ethical theory. The second challenge concerns whether the focus on particular ethical theories (especially rationalist and codifiable ones) in the field of machine ethics involves a biased representation of ethical traditions that contradicts commitments to inclusion and pluralism. In the second part of the paper, we offer a possible candidate for machine ethics we call the "skillful-expert model." This model is based on the assumption that, for machines that have well-defined functions, the requirement of ethical goodness is equivalent to the requirement of *skillful performance*. We argue that this model has strong support from the neglected, virtue-ethical and phenomenological traditions of ethics, and sidesteps the major challenges raised in the first part.

Keywords Machine ethics · Machine virtues · Skill · Moral disagreement · Moral hypocrisy · Philosophy of A.I

F. S. H. Yeung (✉)
School of Philosophical, Historical and Interdisciplinary Studies, University of Essex, Colchester, UK
e-mail: fjwyphil@gmail.com

F. Song
Department of Philosophy, Hong Kong Catastrophic Risk Centre, Lingnan University, Tuen Mun, Hong Kong
e-mail: feisong@ln.edu.hk

L. Checketts, B. S. B. Chan (eds.), *Social and Ethical Considerations of AI in East Asia and Beyond*, Philosophy of Engineering and Technology 47,
https://doi.org/10.1007/978-3-031-77857-5_10

10.1 Introduction: The Two Dogmas of Machine Ethics

This paper has two aims: First, in section two, we seek to critically question the optimistic and rationalist assumptions that underlie a great volume of current research in ethics for machines, and argue that machine ethics should actively engage the problem of moral and social disagreements. Second, we propose a novel model of machine ethics we call the "skillful-expert model" that not only responds to the said challenges in the first part but is also attractive in its own right.

Although the idea of autonomous machines such as self-driving cars and care robots became viable in the past decade, dogmas and first principles are already forming in the terrain of ethics for autonomous machines. One of the ideas commanding the broadest consensus is the *urgency thesis*, which suggests that under the rapid development and deployment of artificial intelligence in increasing domains of life, the development of algorithms that guide machines in making ethically desirable decisions is *necessary* to prevent faulty machines from producing moral disasters.

Closely connected with this, though not a necessary corollary of the first, is the *rationalist thesis* that is often associated with "top-down" approaches to machine ethics. The thesis states that the task for the ethics for (autonomous) machines involves *the identification of explicit principles guiding our moral judgements and the programming of them into various machines.*

10.1.1 Limitations of Non-rationalist Ethics for Machines

Although the urgency thesis does not imply the rationalist thesis, the rationalist thesis is no less important in research in machine ethics. There are several reasons why researchers support this thesis.

For some, the rationalistic assumption follows from reservations about the limitation of alternative approaches to machine ethics. For example, Moor's seminal essay (2006) considers explicit ethical principles necessary for moral machines because machines cannot reliably learn the moral sensitivities and virtues requisite for human-level moral processing. Without reliable ways of training morally sensitive machines, Moor argues that the realistic goal of developing ethical machines should be based on explicit moral algorithms.

Aside from not having capacities essential for full moral agency, people worry that machine learning techniques used for programming non-rationalist theories of morality can lead to problems of untrustworthiness. Machine learning algorithms are known to be "black boxes" whose operations are non-transparent and are prone to non-human-like errors when trained inadequately (Mitchell, 2019). This uncertainty makes error-proofing difficult and can lead to undesirable decision outcomes that pose undue and significant risks to humans (Song & Yeung, 2022; Wallach & Allen, 2009). Besides, training machines in a non-principle-based manner through machine -learning requires large sets of reliable training data. However, there is a question of *what* counts as good moral training data. Empirical data about people's moral responses are available (such as those of people's responses to trolley problems (See, e.g., Awad et al., 2018)), but there is a question of whether they cover the appropriate contexts to produce a program that can provide adequate guidance for autonomous machines. For instance, Goodall (2016, p. 814) argues that the trolley problem "represents a fairly narrow area of automated vehicle ethics and suffers from a perceived lack of realism," and argues that a realistic ethics for autonomous vehicles will need to take into account more nuanced aspects of risks. Also, human responses to moral problems can be heavily tainted by biases and inconsistencies (Baron & Hershey, 1988; Frank et al., 2019), meaning that a workable ethics for machines would require us to screen out such biases in the data.

10.1.2 Positive Reasons for Rationalism

Of course, perceived limitations of non-rationalistic approaches to machine ethics are not the only reasons for supporting rationalist ethical programs. A great number of proposals either support or assume the rationalist paradigm approvingly (E.g., Leben, 2018; Goodall, 2016; Pereira & Saptawijaya, 2011; Susan Leigh Anderson & Anderson, 2011; Arkoudas et al., 2005; Song & Yeung, 2022; Bench-Capon, 2020; Cloos, 2005; Michael Anderson et al., 2006; Powers, 2011). For one of the most radical proposals of the rationalist camp, consider that of the prominent machine ethicist Anderson (2011). Anderson makes the bold proposal that machines can *outperform* humans in ethics, because they are more rational and will not be hindered by cognitive and emotional "biases" befalling the regrettable human mind. Making important assumptions that morality is rationalist, and machines are more rational than human beings, Anderson wants humans to learn ethics from machines.

That most proposals in machine ethics unquestioningly assume a rationalist outlook is hardly accidental. On philosophy's side, rationalist assumptions have for centuries dominated Western philosophy (Dreyfus, 1990, 1992). The terrain of contemporary practical ethics and political philosophy was set amongst rationalists debating which form of rationalism—consequentialist or deontologist—is the right one. Virtue, critical or care perspectives in ethics, to this date, remain marginal and serve as mere critical outsiders in the rationalist terrain.

On the side of engineers, rationalism takes an even stronger root. Engineers are trained as problem-solvers in a strictly rationalist (perhaps even mathematicalizable) sense—problem-solving in engineering often requires unambiguous delineation of concepts and standards of evaluation. Under the current hype surrounding AI and machine learning, there is a tendency for engineers and tech enthusiasts to transport technocratic mindsets and solutions to everything including machine ethics.[1] When a rationalistically disposed philosophy converges with engineering and programming in machine ethics, the rationalist assumption almost naturally becomes the *default position* in the field.

10.2 Challenges to the Rationalist Paradigm

Whichever view one takes about the two assumptions, great volumes of research funding and effort are now channeled into projects investigating machine ethics in a rationalist direction.[2] Given the well-established nature of such an assumption, critical questions about them are not often raised.

10.2.1 Technical Challenges

Many of the problems raised in machine ethics literature are technical. For those who think that machine ethics implementation should take the form of a deterministic algorithm engineered by experts, there is a great strain on the programmer to translate ethical principles into machine-enforceable decision-making procedures that can apply to all relevant situations. For deontologists, this is no easy task, as the classical case in Asimov (2004) shows, even a system comprising three intuitive principles can get into dilemmas in certain situations. The problem only magnifies as more principles and intuitions are programmed—as to what kinds of duties preside over what others, when, and how to translate them into machine-executable language (Allen et al., 2000).

Core consequentialist approaches also face important challenges. For instance, while utilitarian approaches may offer a much more straightforward principle, the challenge is in how an autonomous machine can reliably estimate the utility values of a vast number of scenarios and select the best one. The idea is simple, but when it comes to application, the computational costs may be unrealistically high (Wallach & Allen, 2009, ch. 6; Allen et al., 2000). Besides, how can the machine gather the right sort of information to evaluate all potential outcomes for their decisions? What

---

[1] Consider Ricaurte's (2022, pp. 728–729) remark: "Ethical debates" (as in discussions on AI ethics and regulation) "reflect the underlying assumptions of hegemonic techno-solutionist AI narratives and imaginaries: AI is desirable, AI is unavoidable and AI works (or should work)."

[2] It should be noted that principle-based, rationalist theories are not restricted to strictly deontological and consequentialist ethics, and most rationalist approaches will also incorporate some elements of machine-learning. For instance, Anderson and Anderson (2011) argue for the use of machine-learning to help identify moral principles to be programmed into machines. Bench-Capon (2020) interprets virtues in a way that treats conformity to formally defined virtues as principles for action. Song and Yeung (2022) are agnostic with respect to the validity of the various ethical principles, but remain committed to a principle-based approach to machine ethics.

happens if certain information required for decision-making is unavailable, or only available in low quality? (Song & Yeung, 2022).

It must of course be noted that philosophers are now working with engineers to develop alternatives to the classic theories of machine ethics (See, e.g., Song & Yeung, 2022; Arkoudas et al., 2005), and there may be good reasons to think that technical challenges to rationalist solutions are not insurmountable. However, even when such *technical* hurdles can be overcome, there are more fundamental challenges facing machine ethics.

10.2.2 The "Moral Hypocrisy" Problem

One important problem in the machine ethics literature that has not been given due attention is the problem of "moral hypocrisy" as proposed by Tonkens (2012). Moral hypocrisy describes a situation where the act of developing machines programmed with ethical theory X contradicts theory X. To put it in another way: it may well be the case that some ethical theory is a good ethical theory and can be technically implemented by machines given enough time and resources. However, this alone still does not mean that such a model should be adopted, for as Tonkens (2012) argues, the act of developing moral agents according to a particular theory of ethics implies, at least, the programmer's subscription to that theory. This means that the developer's should also refrain from actions that contradict the theory in question. As Tonken argues, it would be "hypocritical to ask machines to follow moral rules that do not permit their creation in the first place." (140)

Tonkens describes some examples of such hypocrisy in action, including how the development of autonomous weapons may contradict proponents of Asimov's laws of robotics, and proposes that the development of virtuous robots would require us to ask whether it is something that a virtuous engineer would do. In addition to Tonken's examples, we suggest there are even more tricky questions if the social implications of developing ethically intelligent autonomous machines are considered: For instance, we may agree with Leben (2018) that Rawlsian contractualism provides the best ethical theory for machines, but before we consider Rawlsian algorithms, it may be important to figure out *whether* a society that allows the unfettered, capital-driven development of autonomous machines is consistent with norms of justice, and whether such arrangements will be agreed upon by people behind the veil of ignorance.

Consider also how, with the right technical innovations, it may be possible one day to build reliable utilitarian robots. Yet, before this happens, we may have to ask, in line with Singer (1997), whether we should spend huge volumes of capital on AI research, when the same spending can do more to alleviate global hunger and poverty. Alternatively, proponents of climate justice may focus on how autonomous vehicles should have fuel-saving designs, but, given the environmental costs now needed to support the infrastructure for AI development,[3] it might be reasonable to propose to halt the development of AI from the same environmental ethics perspective.

To the extent that the moral hypocrisy problem is a valid concern, the sharp division of labour between those researching on ethics *for (*autonomous) machines and those studying the ethics (and justice) *of* (autonomous) machines should be rethought. If programming ethics for machines is not to fall victim to a hypocritical performative contradiction, it must also engage with the ethical and social challenges that developing such machines poses. *In the extreme cases where the chosen ethical theory clearly prohibits the development of morally autonomous machines, the right response to the problem of machine ethics would not be to go along with the technology and create ethical robots, but to be resolutely against the development of said machines.*

10.2.3 Rationalism and the Fundamental Problem of Ethics

While the moral hypocrisy problem is concerned with the *consistency* between the ethical program and the act of programming such machines, it questions neither the actual validity of any particular ethical theory, nor the validity of the rationalist assumption. However, these latter questions are also important points of contention—as neither can we find reliable consensus around the validity of any particular moral theory, nor is the rationalist assumption as unquestionable as it seems.

If we take a step back, the most fundamental question about both ethics and machine ethics is not whether we can make an agent's actions consistent with a theory's prescriptions, but the question of *what ethics means*, and relatedly, *what values we reflexively deem valuable in humans as in machines*. This very question goes to the heart of every ethical inquiry. Rationalist theories are strongly committed to the *universality, consistency, and rational justifiability* of moral principles. However, the commitment to these values often comes to the exclusion of alternative values that may be equally but not more important.

Even if we limit our scope to the Western tradition in the past century, critiques of rationalism abound. Dancy (2004) offers a critique of principle-based ethics, arguing that moral situations "we encounter differ from each other in subtle ways that no panoply of principles could ever manage to capture." (p. 2) Williams (1985) argues that philosophical morality's insistence on coherence and general obligations misses significant nuances in moral phenomenologies including the feelings of regret and conflicts.

In the "Continental" critical theory tradition, Theodor Adorno argues that rationalist ethics operates within a cold, detached conception of rationality in the philosophical tradition that was responsible for modern moral atrocities such as the

[3] https://www.cbsnews.com/news/artificial-intelligence-carbon-footprint-climate-change/

Holocaust. For Adorno, rationalist ethics hollows out the valuable corporeal and phenomenological aspects of our moral experience (Bernstein, 2001). These aspects—including our moral sensitivities, our repulsion to pain, impulsive "abhorrence" at the senseless suffering of others, and our intuitions about the wrongness of torture—are things a rationalist set of principles fail to capture, but are necessary for an ethics (Adorno, 1975, 2000).

Echoing Adorno in the phenomenological tradition, Levinas (1996) believes that the gist of ethics lies not in the coherence between its precepts, but in our first-hand experience confronting the suffering "face" of the other. The "precariousness" of the Other's "face," for Levinas, is the origin of the prohibition against murder, the source of our responsibility to others, and the "whole gravity of [our] love of the neighbour." Ethics, for Levinas, is not one of figuring out a consistent set of principles and applying them correctly. In sharp contrast with the utopian vision of Anderson (2011), Levinas wrote the following in 1984:

> "It seemed to us important to recall peace and justice as their origin, justification, and measure; to recall that this justice, which can legitimate them ethically... is not a natural and anonymous legality governing the human masses.... Nothing would be able to withdraw itself from the control of the [ethical] responsibility of the "one for the other," which [...]does not cease to appeal to the vigilance of persons who would not be satisfied with the simple subsumption of cases under a general rule, of which a computer is capable." (Levinas, 1996, p. 169)

Against the rationalist view that treats ethics and justice as the application general rules to particular cases (which Levinas disparagingly deems the "computer" to be capable of), Levinas argues that it is the experience with the other that gives any moral imperative and norms of justice its true weight and legitimation.[4]

Moving beyond the terrain of modern, male-dominated Western philosophical traditions, we find other interesting views. In non-Western contexts, rationalist ethics is hardly dominant. Slote (2010) [2013] argued that, in the face of an increasingly globalized world, virtue ethics may deserve more attention. Showing the limitations to the general outlook of contemporary Anglo-American moral philosophy, Slote remarks that "most ethics outside the West has arguably been virtue-ethical in character and that non-virtue-ethical thinking has largely been confined to Western philosophy."

Centuries ago Aristotle disagreed that ethics consists simply of right acts guided by the right principles. Ethics is performed by an agent who has inculcated in themselves the right dispositions to be able to "think, *feel*, and act" in the right way. (Aristotle, 1999 [~340 BCE]) This view is also embraced by contemporary theorists of virtue, who proposed that virtues are more analogous to human *skills*—as also

[4] See also Horkheimer and Adorno (2002 [1969], p. 19) where they question the unreflectiveness of mathematicalized thinking: "Thought is reified as an autonomous, automatic process, aping the machine it has itself produced, so that it can finally be replaced by the machine. Enlightenment pushed aside the classical demand to "think thinking" [...] Mathematical procedure became a kind of ritual of thought. Despite its axiomatic self-limitation, it installed itself as necessary and objective: mathematics made thought into a thing-a tool, to use its own term."

requiring *practice* and sensitivity to context-relevant features that may not be fully capturable in a rational set of principles (Annas, 1995, 2011; Dreyfus & Dreyfus, 2004; Stichter, 2021). In the Chinese tradition, Confucians strongly emphasize moral *education* in shaping moral persons (Wang, 2004). The Daoist philosopher Zhuangzi glorifies the spontaneous and non-intrusive action of actors of skill as an ethical ideal (Ivanhoe, 1993; Zhuangzi, 1891 [350 BCE – 250 BCE]). In Buddhism as well, the way of the Boddhisatva is one of compassion and insight, rather than principles and consistency. Interestingly, both Daoism and Buddhism embrace ethical ideals that are themselves ineffable and can only be practiced (Garfield & Priest, 2021).

Feminist and care-perspectives to ethics also question the rationalist assumptions. Gilligan (1982) strongly questioned the assumptions that ethics should solely be based on abstract, de-contextualized reasonings. Based on her research on women subjects, she discloses a moral-developmental path based on "an injunction to care" as an alternative to one based on abstract, decontextualized conceptions of rights and principles (p.100). Going a more radical path, Nodding (1984, ch. 4) argues that ethics as an "active virtue" cannot be severed from our natural dispositions to care and our response to the suffering of others.

We mentioned the various critiques and alternatives to the dominant rationalist one so as to reopen the field of machine ethics to important questions and alternatives it needs to engage. Instead of working with the tunnel vision that ignores all theoretical disagreements, machine ethics should be aware of the uncertain *ethical basis* it treads on (Brundage, 2014). As a branch of ethics, machine ethics reproduces long-standing controversies and disagreements in the realm of ethics. Even as philosophers are trying to rationally reflect on their respective justifications, consensus is nowhere to be found. The age-old question of whether deontology or consequentialism gives the right ethical principles remains unresolved, as challenges come from critical theories, pheonomenologies, virtue ethics or ethics of care. More fundamentally the theoretical question involves whether ethical rationalism is defensible, whether affective sensitivity is necessary for ethics, and whether ethics can ever be articulated into principles.

Against the rationalist tendency, we should not regard all disagreements as results of deficient rationality. Ethical disagreements can legitimately come from differences in core value commitments resulting from different intuitions, cultural traditions, and (gendered) experiences. Indeed, even if an ethical theory can command consensus within a large group of researchers, there remains the social and *political* question of whether it is justified for such 'experts' to impose the consensus on the general public through algorithmic design. We are living in post-traditional societies defined by high degrees of pluralism in beliefs and values (Rawls, 2005; Habermas, 1992). When society-wide decisions are made on the ethical design of machines that has significant impact on the general public—we want these decisions to be made in ways that are consistent with the democratic principle of respecting the interests and value orientations of everyone.[5] Even if we do not aim to forge a society-wide consensus surrounding machine ethics, the general public should have a choice to disagree and a right to veto which ethic to adopt (Bonnefon et al., 2020; Millar, 2015, 2017).[6] All in all, machine ethicists should be aware that, unlike engineering problems, ethical solutions do not and should not come "top-down" from the "reflective" minds of expert ethicists and engineers. If machine ethics is not to fall into dangers of elitism and technocraticism, the facts of ethical pluralism and the ineradicability of disagreements should be respected in the design of ethical machines.

10.3 Going Forward: Virtuous Machines?

As we have seen, a strong alterative to rationalist ethics is virtue ethics. Virtue ethical theories are present in various ethical and cultural traditions. Instead of applying rules and principles, schools of virtue ethics focus on how an agent's virtuous character (pre-reflectively and affectively but reliably) *disposes* them to respond (virtuously) to various situations. Indeed, virtue theories can see compliance to rigid principles as a moral deficiency, as virtuous responses should be highly adaptive and sensitive to the particularities of novel situations. Relatedly, virtuous characters and dispositions are acquired like skills—i.e. *learned and refined* through repeated exposure to exemplars and practice (Annas, 1995; Stichter, 2007, 2013). Inspired by these new theories, a few notable attempts in the last decade have tried to adopt bottom-up machine learning methods to produce virtuous behaviour in machines (Howard & Muntean, 2016, 2017; Stenseke, 2021, 2022; Berberich & Diepold, 2018; Gamez et al., 2020; Wallach & Allen, 2009).

These approaches are welcome alternatives to rationalist ones. However, they also produce their own problems. For instance, a virtue-ethical machine ethic that tries to incorporate affective and situational sensitivity may be an advantage over its rationalist counterparts, but the same advantage may also make programming such machines either difficult or morally hypocritical. Importantly, there is a question of whether machines can exhibit virtuous character when there are widespread doubts about whether machines can possess characteristics that define the virtuous character of human agents. This includes highly complex systems that have to model emotions, phenomenological sensitivity, consciousness, intentionality and the ability to

[5] In referring to the "democratic principle" I am extending the discussion in Jürgen Habermas (1996). For a proposal to "democratize innovation" in AI, see Sadowski (2020, ch.8).

[6] Bonnefon et al.'s argument is that the social benefits of expert-designed moral algorithms may lose its social benefits if laypersons prefer to *opt out* because they do not see the reason to endorse these algorithms. Indeed, the problem might be deeper—the expert's imposition of moral values through algorithms may well count as a form of *epistemic injustice* against the general public, when their disagreements and value orientations are not well represented in the production of moral algorithms bearing widespread social effects.

understand other minds (Tonkens, 2012, p. 142; Railton, 2020; Howard & Muntean, 2017, pp. 127–8).

For instance, Wallach and Allen (2009) proposed to adopt the "learning intelligent distribution agent" (LIDA) architecture to model the complex interactions between affects and conscious moral deliberations in a human-like manner. This project is highly ambitious technically, but the challenge of moral hypocrisy looms if it succeeds. Indeed, as Tonkens (2012) argues, if such machines capable of high degree of (moral) autonomy are created, there is a question of whether we would be "disrespecting the autonomy of an otherwise autonomous moral agent, [...] limiting the ways that such agents can express their autonomy." (146) Designing such machines may bespeak moral hypocrisy, as it may be inconsistent with *virtuous character* for any developer to create machines with such affective capacities.

Besides, the *disagreement* problem still exists for virtues. Although programming virtue ethics does not require us to seek programmable principles, the selection of virtuous modes of action in turn also draws upon a certain understanding of what the good is. This is in operation in virtue ethics involve the selection of moral exemplars and "eudaimonistic functions" (See, e.g., Stenseke, 2021, 2022). This inevitably requires the programmer to decide *which human ends are worth pursuing* and which are less worthy. In our world of ineliminable cultural and moral pluralism, this simply can hardly be done without reasonable room for disagreements.

10.4 The Skillful-Expert Model for Machine Virtues

To proceed, we propose a skillful-expert model of machine virtues as a possible answer to the ethics of machine challenges. To introduce our skillful-expertise model, this section tries to show how difficult discussions can be sidestepped in machine ethics by (i) considering ethics in relation to the functions machines themselves are designed to serve and comparing how humans handle similar functions, and then (ii) proposing to develop machines that reproduce and mimic human skills in such domains.

To begin with, let's look at practices such as driving and healthcare. These two practices are apt since these are domains where autonomous machines are actively being developed. When we look at how humans deal with driving and healthcare, human drivers and nurses seldom busy themselves with figuring out the right model of ethics and applying them constantly: unlike consequentialist robots, they do not (nor are they able to) constantly calculate risks for all likely courses of actions; and unlike deontological robots, their decisions are hardly determined by principle-based justifications.[7] Yet, despite such "failures," we seldom consider drivers and nurses significant moral hazards and have little reason to do so if the experts in

[7] This is only to argue that most decisions made under skill are often not motivated by considered judgements that refer to explicit rules and principles. It might be true, especially for decisions that can potentially general high risks, that decisions should be retrospectively justifiable—as, in cases

question have a good track-record. This is because, although experts do not follow principles, they have *skills* acquired through adequate training and practice to help them handle their daily tasks and mitigate risks to people around them. In the human context, rules and guidelines do exist to guide people's actions and (especially) prohibit clear instances of wrong-doing. However, these rules seldom dictate every decision they make. This is because people's skills, acquired from experience, reliably guarantee the goodness of outcomes and are the basis of our trust in the experts.

For machine ethicists under the rationalist tradition, the morality of decisions made by autonomous machines and, relatedly, their trustworthiness depends on *how well machines apply ethical principles and explain their decisions.* Naturally, machines, they believed, should be programmed with explicit principles to guide their decisions. However, if we compare machine performance to human performance, the assumption that machines can only be trustworthy and moral when principles guide their actions seems problematic. To be sure, we do not want our drivers and doctors to be brash and disregard ethical codes of conduct altogether, but it would equally be questionable if our laws requires require every doctor and driver to measure every decision they make against set standards, principles, and protocols. Indeed, we would be less trustful of doctors and drivers who do this, as it is an indication that *they lack the right skill.*

Hubert Dreyfus and Stuart Dreyfus provide an interesting model of skill which confirms our intuitions about experts and challenges the idea that intelligent actions mean deciding on the basis of principles (Dreyfus, 2004; Dreyfus & Dreyfus, 1986). In their account, a novice acquires skills first by mechanically following and applying context-dependent rules appropriate to that practice. This description matches what rationalists machine ethicists envision what ethical machines should be like. However, for humans, this is just the *beginning* of the journey.

With sufficient practice, human agents start to become *familiar* with the skill-context. This familiarity is reflected in being able to respond *intuitively and spontaneously (without much thought) to an increasing number of situations.* This growing ability to unconsciously respond to aspects of the situation intuitively and spontaneously allows the competent learner to further refine their skills by focusing on aspects of the situation that they have not yet mastered and decide explicitly what to do. These deliberate decisions can create varying results, and the expert-to-be will perceive such results as feedback to improve on the skill. As the competent learner continues to practice, even those situations that were troubling become more *intuitive*, and their activity reaches a state that converges with what recent psychologists have termed "flow" (Nakamura & Csikszentmihalyi, 2002). In the words of Dreyfus, the expert possesses "intuitions" to help them *'see'* and *'feel'* the right response at any moment. Describing the ideal state of the expert, Stuart Dreyfus writes:

> "The expert not only sees what needs to be achieved... he or she also sees *immediately* how to achieve this goal.... With enough experience in a variety of situations, all seen from the

---

where accidents do happen, the expert would and should be held accountable for the decisions they made.

> same perspective but requiring different tactical decisions, the brain of the expert gradually decomposes this class of situations into subclasses, each of which requires a specific response. This allows the immediate intuitive situational response that is characteristic of expertise." (Stuart E. Dreyfus, 2004, pp. 179–180, emphasis added)

There are two interesting points about Dreyfus's description: First, Dreyfus's description of "decomposing [classes] of situations into subclasses" and then connecting them to viable solutions are remarkably similar to complex classification problems deep machine-learning algorithms are good at emulating. It is, therefore, interesting to explore whether this goal of skillful expertise can be reproduced by machines trained with data gathered from experts.

Second, and more importantly, despite reservations we may have towards Dreyfus's extreme emphasis on the automatism of expertise (See, Stichter, 2018, pp. 48–52), his account does point to a strong alternative to the rationalistic approaches in ethics for machines. If human expert decisions are (at least to an important extent) automatic and non-principle based, but at the same time this automatism is not a sufficient reason for us to distrust human experts, then it makes sense for us to at least be open to the possibility of considering non-principle based *machine* experts, and potentially trusting them when their performance is on par with experts.

These are key motivations for developing the skillful-expert model of machine ethics. Under this model, we suggest that machine ethicists can at least partly refrain from adopting principle-based programs and seek to emulate the unconscious, intuition-driven aspect of skills through machine-learning methods.

10.4.1 Philosophical Support for the Skillful-Expertise Model: Skills as Machine Virtues

At this point, one might respond that how the machine performs in everyday situations does not fall into the domain of ethics. Thus, skills do not fall into the domain of machine ethics but, perhaps, a question of safety and responsible design to be left to engineers. There are two reasons for us to consider skills as a potential candidate for machine ethics, one pragmatic and the other philosophical.

First, while we may not usually think of skills themselves as ethical, the results they produce can be morally evaluated. Although not explicitly risk calculating, skillful actions effectively negotiate between potentially conflicting human values (in the case of driving, they may include passenger comfort, road safety, fuel efficiency). Besides, decisions made by experts are likely to be more context-sensitive and therefore much less risky and accident-prone than actions done by unskilled agents rigidly applying rules. For instance, autonomous vehicles with unskillful behaviour can greatly increase the risk of crashes, while one that is context-aware and responsive skillfully is likely to avoid accidents.[8] This ability to reduce risk to human agents is itself a morally desirable feature of skillful performance. Besides, to the degree that some human practices[9]—as a whole—contribute to general human good (e.g. good medical practices, good driving, etc.) and that skillful activities in these practices guarantee the optimal functioning of these practices, we can say that such skills can (at least indirectly) contribute to human flourishing.

Second, there are significant support for skills as virtues in the philosophical literature. As we have mentioned in Sect. 10.2.3, the Chinese philosopher Zhuangzi has proposed that skillful actions are the model for human excellence. In various passages, he proposed that the actions of the skillful swimmer, butcher, and wheelwright, exemplify "de" (virtue) through the love of "Dao." What Zhuangzi argue in these passages is that there can be multiple Daos, and what the Dao requires for different virtuous agents in different contexts depends on how well the agent in question performs a harmonious and non-instructive course of action that accords with the Dao of its environment and other agents. (See, Ivanhoe, 1993) Indeed, Zhuangzi's contextual and non-anthropocentric conception of skills as virtue can also apply to *machines*, as the same skill is what may bring about harmony in the respective practices.

Besides, there is an Aristotelian argument for the ethicality of machine skills. We have in Sect. 10.3 argued that existing virtue-ethical accounts of machine ethics face important challenges because they try to transpose human virtues onto machines. However, this is not the only way the virtue-ethical approach can be applied to machine ethics. For example, in Book II of the *Nicomachean Ethics*, Aristotle (1999 [~340 BCE]) argues that the good for any practice and agent depends on what the characteristic functions it is supposed to perform, and that virtues allows the agent to perform these functions well (Korsgaard, 2008). Aristotle defines a human being's characteristic activity as *rationality* (logos) and *ultimate end* as *eudaimonia*. But when it comes to machines, we may need different ends and conceptions of functioning. As Coleman (2001, p. 248) suggests, "virtues must be identified in relation to the individual [or agent] whose excellence is in question." For most machines, their end is often technically and instrumentally defined, this means that our conception of machines' excellence and virtuousness should also issue from such an understanding of ends and functioning. In the skillful-expert model, we affirm that the ends of machines are *technically* defined, and for machines working in human practices where skills have a place, the corresponding "virtues" for such machines should also be modelled like human skills. In other words, *skills* can—following Aristotle's arguments—be "machine virtues."

[8] Perhaps following Goodall (2014, p. 97)'s proposal, we should consider "any activity that transfers risk from one person to another [to involve] ethics"—at least in the domain of machine ethics.

[9] Certainly not *all* human practices requiring skills fall into this category. For instance, the skill of murder would not be.

10.4.2 Skills, Moral Hypocrisy and Disagreement Problem

Before concluding, we shall return to how skillful expert model deals with the moral hypocrisy and disagreement problems outlined in Sect. 10.2.

As it should be apparent by now, the skillful-expert model carries much thinner ethical commitments than most existing theories in machine ethics. By treating machine virtues as analogous to human skills, we are mainly committed to four ideas:

- (i) There are social practices in which the use of well-functioning machines may bring about human good overall,
- (ii) For at least some of these skill domains listed in (i), an agent can perform its task skillfully or unskillfully.
- (iii) For domains where both (i) and (ii) are satisfied, it is good for the machine to perform skillfully rather than unskillfully.
- (iv) Good performance of machines can be emulated if machines are trained to mimic the actions of human experts.

It is hard to see how moral hypocrisy can be a systematic problem for ethicists who want to design skillful machines. The skilful model is intended to be an ethical theory for machines, not humans.[10] Indeed, the only apparent moral contradiction that can render proponents of the skillful-expert model for machines hypocritical is when they do not commit themselves to developing their skills in programming. So long as they can readily assume that machine ethicists would want to do well and be skillful at doing what they do (why even do it if they are not committed to doing it well?), this requirement is met.

Concerning the disagreement problem, we may understand our model as a non-intrusive and non-contentious way of integrating machines into existing human practices, insofar as such approach seeks to reproduce the *unproblematic* practices of humans as far as possible, and not to seek to *introduce/program* something anew. Regarding the core assumptions listed above, it seems that (ii) and (iii) can hardly be rejected, and disagreements about (iv) only revolve around the technical feasibility of skillful machines and it is not itself a moral problem. Therefore, most ethical controversies are likely to surround (i).

We admit that there are a number of practices where people disagree as to whether their optimal functioning would bring about human good (e.g. autonomous weapons and trading robots). However, there also seems to be some practices (e.g. good patient care, safe driving) where the connection between the practice and human good aligns. Within these practices, it is also not difficult to find consensus around who are the experts with established and good rapport. Therefore, disagreements about (i) would not be prevalent.

---

[10] In arguing for the skillful-expert model of machine ethics, the model is agnostic as to which theory of *human* ethics is the right theory. Our model only claims that skills are virtues for autonomous *machines* with specific, well-defined functions. Claiming this does not entail claiming that skills themselves qualify as *human* virtues.

10.5 Conclusions

We are now approaching an age where autonomous machines are going to perform increasingly important functions in society. If we want these machines to perform ethically, one uncontentious solution might not be to program them to follow explicit moral codes, but doing so can invite challenges of moral hypocrisy and disagreements. The skillful-expert model proposes that the machine-ethics question can potentially be answered by designing machines that can emulate human skills in the practices they are designed to perform. If we cannot find uncontentious solution to trolley problems, at least we may be able to design good autonomous cars that can skillfully behave well on the road and avoid crashes.

References

Adorno, T. W. (1975). *Negative Dialektik*. Suhrkamp.

Adorno, T. W. (2000). *Problems of moral philosophy* (R. Livingstone, Trans.). Polity Press.

Allen, C., Varner, G., & Zinser, J. (2000). Prolegomena to any future artificial moral agent. *Journal of Experimental & Theoretical Artificial Intelligence, 12*(3), 251–261. https://doi.org/10.1080/09528130050111428

Anderson, S. L. (2011). How machines might help us achieve breakthroughs in ethical theory and inspire us to behave better. In M. Anderson & S. L. Anderson (Eds.), *Machine ethics* (pp. 524–530). Cambridge University Press.

Anderson, S. L., & Anderson, M. (2011). A prima facie duty approach to machine ethics: Machine learning of features of ethical dilemmas, prima facie duties, and decision principles through a dialogue with ethicists. In M. Anderson & S. L. Anderson (Eds.), *Machine ethics* (pp. 476–492). Cambridge University Press.

Anderson, M., Anderson, S. L., & Armen, C. (2006). An approach to computing ethics. *IEEE Intelligent Systems, 21*(4), 56–63.

Annas, J. (1995). Virtue as a Skill. *International Journal of Philosophical Studies, 3*(2), 227–243. https://doi.org/10.1080/09672559508570812

Annas, J. (2011). *Intelligent Virtue*. Oxford University Press.

Aristotle. (1999 [~340BCE]). *Nicomachean Ethics* Edited and translated by T. Irwin. Hackett.

Arkoudas, K., Bringsjord, S., & Bello, P. (2005). Toward ethical robots via mechanized deontic logic. In *AAAI fall symposium on machine ethics* (pp. 17–23). The AAAI Press Menlo Park, CA, USA.

Asimov, I. (2004). *I, Robot*. Bantam Dell.

Awad, E., Dsouza, S., Kim, R., Schulz, J., Henrich, J., Shariff, A., et al. (2018). The moral machine experiment. *Nature, 563*(7729), 59–64. https://doi.org/10.1038/s41586-018-0637-6

Baron, J., & Hershey, J. C. (1988). Outcome bias in decision evaluation. *Journal of Personality and Social Psychology, 54*(4), 569–579.

Bench-Capon, T. J. M. (2020). Ethical approaches and autonomous systems. *Artificial Intelligence, 281*, 103239. https://doi.org/10.1016/j.artint.2020.103239

Berberich, N., & Diepold, K. (2018). *The virtuous machine – old ethics for new technology?* ArXiv. https://doi.org/10.48550/arxiv.1806.10322

Bernstein, J. M. (2001). *Adorno: Disenchantment and ethics*. Cambridge University Press.

Bonnefon, J.-F., Shariff, A., & Rahwan, I. (2020). The moral psychology of Ai and the ethical opt-out problem. In S. M. Liao (Ed.), *Ethics of artificial intelligence*. Oxford University Press.

Brundage, M. (2014). Limitations and risks of machine ethics. *Journal of Experimental & Theoretical Artificial Intelligence, 26*(3), 355–372. https://doi.org/10.1080/0952813X.2014.895108

Cloos, C. (2005). The utilibot project: An autonomous mobile robot based on utilitarianism. In *2005 AAAI fall symposium on machine ethics* (pp. 38–45).

Coleman, K. G. (2001). Android Arete: Toward a virtue ethic for computational agents. *Ethics and Information Technology, 3*(4), 247. https://doi.org/10.1023/A:1013805017161

Dancy, J. (2004). *Ethics without principles*. Oxford University Press.

Dreyfus, H. L. (1990). Is Socrates to blame for cognitivism? In B. Goeranzon & M. Florin (Eds.), *Artificial intelligence, culture and language: On education and work* (pp. 219–228). Springer.

Dreyfus, H. L. (1992). *What computers still can't do: A critique of artificial reason*. MIT Press.

Dreyfus, S. E. (2004). The five-stage model of adult skill acquisition. *Bulletin of Science, Technology & Society, 24*(3), 177–181. https://doi.org/10.1177/0270467604264992

Dreyfus, H. L., & Dreyfus, S. E. (1986). *Mind over machine: The power of human intuition and expertise in the era of the computer*. Free Press.

Dreyfus, H. L., & Dreyfus, S. E. (2004). The ethical implications of the five-stage skill-acquisition model. *Bulletin of Science, Technology & Society, 24*(3), 251–264. https://doi.org/10.1177/0270467604265023

Frank, D.-A., Chrysochou, P., Mitkidis, P., & Ariely, D. (2019). Human decision-making biases in the moral dilemmas of autonomous vehicles. *Scientific Reports, 9*(1), 13080. https://doi.org/10.1038/s41598-019-49411-7

Gamez, P., Shank, D. B., Arnold, C., & North, M. (2020). Artificial virtue: The machine question and perceptions of moral character in artificial moral agents. *AI & SOCIETY, 35*(4), 795–809. https://doi.org/10.1007/s00146-020-00977-1

Garfield, J. L., & Priest, G. (2021). Skill and virtuosity in Buddhist and Daoist philosophy. In E. Fridland & C. Pavese (Eds.), *The Routledge handbook of philosophy of skill and expertise* (pp. 29–39). Routledge.

Giligan, C. (1982). *In a different voice*. Harvard University Press.

Goodall, N. J. (2014). Machine ethics and automated vehicles. In G. Meyer & S. Beiker (Eds.), *Road vehicle automation* (pp. 93–102). Springer.

Goodall, N. J. (2016). Away from trolley problems and toward risk management. *Applied Artificial Intelligence, 30*(8), 810–821. https://doi.org/10.1080/08839514.2016.1229922

Habermas, J. (1992). *Postmetaphysical thinking: Philosophical essays* (W. M. Hohengarten, Trans.). Polity.

Habermas, J. (1996). *Between facts and norms: Contributions to a discourse theory of law and democracy* (W. Rehg, Trans.). MIT Press.

Horkheimer, M., & Adorno, T. W. (2002 [1969]). *Dialectic of enlightenment: Philosophical fragments* (E. Jephcott, Trans.). Stanford University Press.

Howard, D., & Muntean, I. (2016). A minimalist model of the artificial autonomous moral agent (Aama). In *SSS-16 symposium technical reports* (pp. 217–225). Association for the Advancement of Artificial Intelligence, AAAI.

Howard, D., & Muntean, I. (2017). Artificial moral cognition: Moral functionalism and autonomous moral agency. In T. M. Powers (Ed.), *Philosophy and computing: Essays in epistemology, philosophy of mind, logic, and ethics* (pp. 121–160). Springer.

Ivanhoe, P. J. (1993). Zhuangzi on skepticism, skill, and the ineffable Dao. *Journal of the American Academy of Religion, LXI*(4), 639–654. https://doi.org/10.1093/jaarel/LXI.4.639

Korsgaard, C. M. (2008). Aristotle's function argument. In *The constitution of agency: Essays on practical reason and moral psychology* (pp. 129–150). Oxford University Press.

Leben, D. (2018). *Ethics for robots: How to design a moral algorithm*. Routledge.

Levinas, E. (1996). *Emmanuel Levinas: Basic philosophical writings*. ndiana University Press.

Millar, J. (2015). Technology as moral proxy: Autonomy and paternalism by design. *IEEE Technology & Society Magazine, 34*(2), 47–55. https://doi.org/10.1109/MTS.2015.2425612

Millar, J. (2017). Ethics settings for autonomous vehicles. In P. Lin, K. Abney, & R. Jenkins (Eds.), *Robot Ethics 2.0: From autonomous cars to artificial intelligence* (pp. 35–50). Oxford University Press.

Mitchel, M. (2019). *Artificial intelligence: A guide for thinking humans*. Penguin.

Moor, J. H. (2006). The nature, importance, and difficulty of machine ethics. *IEEE Intelligent Systems, 21*(4), 18–21.

Nakamura, J., & Csikszentmihalyi, M. (2002). The concept of flow. In *Handbook of positive psychology* (pp. 89–105). Oxford University Press.

Noddings, N. (1984). *Caring, a feminine approach to ethics & moral education*. University of California Press.

Pereira, L. M., & Saptawijaya, A. (2011). Modeling morality with prospective logic. In M. Anderson & S. L. Anderson (Eds.), *Machine ethics* (pp. 398–421). Cambridge University Press.

Powers, T. M. (2011). Prospects for a Kantian machine. In M. Anderson & S. L. Anderson (Eds.), *Machine ethics* (pp. 464–475). Cambridge University Press.

Railton, P. (2020). Ethical learning, natural and artificial. In S. M. Liao (Ed.), *Ethics of artificial intelligence* (pp. 48–78). Oxford University Press.

Rawls, J. (2005). *Political liberalism*. Columbia University Press.

Ricaurte, P. (2022). Ethics for the majority world: Ai and the question of violence at scale. *Media, Culture & Society, 44*(4), 726–745.

Sadowski, J. (2020). *Too smart : How digital capitalism is extracting data, controlling our lives, and taking over the world*. MIT Press.

Singer, P. (1997). The drowning child and the expanding circle. *New Internationalist*.

Slote, M. (2010). Virtue ethics. In J. Skorupski (Ed.), *The Routledge companion to ethics* (pp. 478–489). Routledge.

Song, F., & Yeung, F. S. H. (2022). A pluralist hybrid model for moral Ais. *AI & Society*. https://doi.org/10.1007/s00146-022-01601-0

Stenseke, J. (2021). Artificial virtuous agents: From theory to machine implementation. *AI & Society*. https://doi.org/10.1007/s00146-021-01325-7

Stenseke, J. (2022). Artificial virtuous agents in a multi-agent tragedy of the commons. *AI & Society*. https://doi.org/10.1007/s00146-022-01569-x

Stichter, M. (2007). Ethical expertise: The skill model of virtue. *Ethical Theory and Moral Practice, 10*(2), 183–194. https://doi.org/10.1007/s10677-006-9054-2

Stichter, M. (2013). Virtues as skills in virtue epistemology. *Journal of Philosophical Research, 38*, 333–348. https://doi.org/10.5840/jpr20133817

Stichter, M. (2018). *The skillfulness of virtue: Improving our moral and epistemic lives*. Cambridge University Press.

Stichter, M. (2021). Virtues as skills, and the virtues of self-regulation. *The Journal of Value Inquiry, 55*(2), 355–369. https://doi.org/10.1007/s10790-021-09820-2

Tonkens, R. (2012). Out of character: On the creation of virtuous machines. *Ethics and Information Technology, 14*(2), 137–149. https://doi.org/10.1007/s10676-012-9290-1

Wallach, W., & Allen, C. (2009). *Moral machines: Teaching robots right from wrong*. Oxford University Press.

Wang, F. (2004). Confucian thinking in traditional moral education: Key ideas and fundamental features. *Journal of Moral Education, 33*(4), 429–447. https://doi.org/10.1080/0305724042000327984

Williams, B. (1985). *Ethics and the limits of philosophy*. Harvard University Press.

Zhuangzi. (1891 [350BCE – 250BCE]). 'The Writings of Chuang Tzu'. Available at: https://ctext.org/zhuangzi.

Chapter 11
Universal Values in AI Ethics

Soraj Hongladarom

Abstract In this chapter, I would like to argue that there are universal values in AI ethics. This line of argument is quite neglected in the voluminous literature in the field. Basically, the argument is that the basic values in AI ethics that are generally agreed on, such as privacy, explainability, lack of bias, and so on, are indeed universalizable. Furthermore, given the empirical situation of the widespread use of AI and the shared global communication systems, these values are indeed universal. The idea, to be argued for in the chapter, is that the basis of this universality is founded not on any *a priori* theorization, but on the contingencies or exigencies of the situation that obtains empirically. The reason for this is that values always serve certain goals, and the main AI values serve goals that happen to be advantageous to those who work with AI as well as those who benefit from it, the advantage that spans cultural differences. This does not necessarily mean that cultures are essentially the same, but it shows that cultures cannot remain in isolation, but always interact with one another, and that this is especially the case when it comes to AI.

Keywords Values · AI ethics · Universality · Culture

11.1 Introduction

Most discussions on AI ethics focus on the normative aspects of the discipline. For example, in a well-known introductory work on the field, Mark Coeckelbergh focuses almost exclusively on the values and normative assessments that stakeholders in AI should follow (Coeckelberg, 2020). He focuses on fields such as superintelligence, transhumanism, the fundamental differences between humans and machines, moral agency and patiency, biased algorithms, privacy, explainability,

S. Hongladarom (✉)
Center for Science, Technology, and Society, Chulalongkorn University, Bangkok, Thailand
e-mail: hsoraj@chula.ac.th

L. Checketts, B. S. B. Chan (eds.), *Social and Ethical Considerations of AI in East Asia and Beyond*, Philosophy of Engineering and Technology 47,
https://doi.org/10.1007/978-3-031-77857-5_11

and so on, but there is little discussion on the more theoretical level. Questions such as why individual privacy should be accorded so much value or what is actually wrong with biases in algorithms and machine learning are given relatively scant attention. Basically, these are questions that focus on the justification of ethical pronouncements, rather than on the pronouncements *per se*. This may be understandable in a short introductory work, which focuses more on what needs to be done and what exactly are thought to be right or wrong in AI ethics, but the problem does not limit itself to Coeckelbergh's work here. A brief survey of the vast literature in the emerging field of AI ethics shows that questions about justification of ethical pronouncements have been given relatively little attention (some exceptions are Bleher & Braun, 2023; Paraman & Anamalah, 2023). It is as if scholars and practitioners in the field already agree on what is right and what is wrong in AI ethics. It seems as if everybody agrees that, for example, violation of individual privacy is wrong, and the task of AI systems is to avoid doing that. However, questions of the justificatory or theoretical kind are still important because they give us an understanding of the reason or rationale underlying the ethical pronouncements. And these reasons or rationales are of interest in themselves. Moreover, as I shall try to show during the course of this chapter, they are also of *practical* significance, especially when it is found that these reasons and rationales can widely differ from one theorist to another, and from one cultural domain to another. As AI technology is spanning the globe, it is bound to be conceptualized in the differing cultural and geographical domains in different ways, especially with regard to ethics and values, which are highly susceptible to varying cultural differences. A fuller understanding of how different cultures view AI and AI ethics thus has a tremendous practical benefit of creating an environment where the truly global functioning of AI is possible. Furthermore, engaging in cross-cultural dialogs on AI ethics could lead to deeper acceptance of certain values across cultural domains too.

I will focus on whether it is possible for there to be universal values in AI ethics. I will argue that there are but there are certain caveats as we shall see in the paper. Discussing universal values in AI ethics seems to be rather neglected in the literature, as most scholars appear to agree that the values associated with AI ethics, e.g., privacy, lack of bias, and explainability, are valid across cultural domains, so much so that literature on non-Western perspectives in AI ethics and on the need for cross-cultural engagement are scarce. My discussion of the universal values in AI ethics is part of a larger project of grounding and justifying AI ethical values on the exigencies of the empirical conditions in various geographical domains. Values, especially those being discussed in AI ethics discourse, are, or should be, those whose upholding leads to the fulfillment of certain goals. These goals can vary across cultures, so can the values. However, in the case of AI ethics, these values happen to be rather similar (see, for example, Hongladarom & Bandasak, 2023). In this case, then, the values are universal on the basis of the general agreement among cultures that these values serve largely the same set of goals.

In the next section, I will conduct a brief survey of the current literature in the field in order to show that most works in AI ethics focus on the practical level of identifying the principles and translating them into actual practice. Section 11.3

discusses some of the reasons why investigating the theoretical foundations of AI ethics is necessary and, perhaps paradoxically, of practical significance. Section 11.4 deals with the central topic of the paper, namely the question of universality of values in AI ethics. Then Sect. 11.5 concludes the paper.

11.2 A Brief Survey of the Literature

AI ethics is a rapidly growing field; its growth has exploded so much that it is very difficult to keep up with the field as a whole. Nonetheless, a few patterns seem to emerge in the last few years of the literature on the topic. Firstly, it is noticeable that most works in the field, especially in the past few years, focus on the *practical* aspects of the ethics (e.g., on how to operationalize AI ethics), so much so that theoretical investigations into the conceptual or normative foundations of AI ethics seems to be rather neglected. And secondly, in the relatively few works that discuss the normative theories, these theories are almost invariably based on philosophical ethics as is practiced in the West. As for the first pattern, among some of the relatively few works that mention theory at all are Huang et al. (2023) and Hagendorff (2022). Huang et al. (2023) is a thorough overview of the literature and the topics in AI ethics as a whole; the amount of literature surveyed in the paper is impressive. They discuss all the three main ethical theories in the West, namely Kantian deontological theory, utilitarian theory, and virtue ethics theory (Huang et al., 2023). Hagendorff (2022) is also a thorough investigation of the issues involved in AI ethics; his paper differs from Huang in that he has a critical perspective, where he argues that the main ethical principles found in most AI ethics guidelines (viz., privacy, explainability, lack of bias—see also Blackman, 2022) are "calculable" and designed only to facilitate the work of the technicians involved in working with AI. What is missing in his view are the sociocultural contexts in which AI is developed and deployed (Hagendorff, 2022). However, neither Huang nor Hagendorff mention anything about the problem of universality of values or principles. It is as if they tacitly believe that these values are universal and applicable everywhere in the world. Another work that comes close to mentioning or discussing the theories is Ashok et al. (2022). In this paper Mona Ashok and her co-authors also conduct a systematic and comprehensive review of works on AI ethics (Ashok et al., 2022), and found 14 areas which most works on AI ethics and ethics of digital technologies focus on. Nonetheless, these 14 areas still do not contain any mention of the problem of the universality of values. Furthermore, recent works tend to focus on how to "operationalize" AI ethics rather than discussing its theoretical foundations. For example, a recent paper by Jessica Morley and her colleagues (Morley et al., 2023) calls for effort to close that gap between theory and practice by attempting to answer questions that are related to how the gap can be closed (Morley et al., 2023). Hannah Bleher and Matthias Braun (Bleher & Braun, 2023) reflect on how AI ethics has been put into practice and propose certain factors that they argue are necessary for successfully translating AI ethics principles into practice. While these attempts are laudable, they do not directly focus on which normative theories are more tenable or how to justify them.

The second pattern that emerged from the brief survey of the literature is that, when these works deal with the ethical principles or theories at all, they almost exclusively focus on Western ethics. In fact, the three main ethical theories that come to the mind of those who have any acquaintance with ethics are all based on ethics as studied and practiced in the West. The few exceptions to this rule are my own works (e.g., Hongladarom, 2016, 2020), works by Peter Hershock (2021) and James Hughes (2012), and an article by Bryce Goodman (2022). In my book (Hongladarom, 2020), I proposed a theory of AI ethics that is based on Buddhist philosophy rather than the familiar philosophy of the West. What is surprising is that, given the fact that AI has become totally pervasive in many non-western countries, discussion in the current literature still focuses almost exclusively on theories originated from the West. This parochialism perhaps explains why discussions on the universality of values are so scarce in the voluminous literature on AI ethics. A textbook on the field by Paula Boddington, for example, discusses in detail the main three ethical theories, but does not mention any other theories besides these three at all (Boddington, 2023).

11.3 Why Discussion of Theories Is Necessary

There is a huge gap in the literature on the non-Western approaches to the normative considerations to ethics, as well as on the treatment of theoretical investigations in the field overall. This gap needs to be filled because AI does not respect any national or cultural boundary, and its impacts do not limit themselves only to the West. Moreover, understanding the non-Western approaches does lead to a deeper understanding of the theoretical foundations of AI ethics as a whole, so focusing on the non-Western approach is not a parochial interest belonging only to specialists.

However, before we focus on the non-Western approaches, there is the important question of why theoretical investigation in AI ethics is necessary in the first place. AI ethics as it is being discussed at this moment is an interdisciplinary field, where scholars from many different disciplines participate. Even when philosophers do join in the discussion, their style of engaging with their colleagues in the field tends to diverge from their style when they talk to their fellow philosophers only. What stands out is that there are many references to empirical research, both in the broad social scientific domain of looking at the social context in which AI is used and in computer science itself. This is a welcome trend, as it opens up philosophy to a wider audience as well as broadening the horizons of philosophers themselves. However, engaging with the social sciences and computer science should not result in philosophers losing sight of their mission, which is to engage in conceptual investigation and searching for theoretical foundation and justification of normative principles. Only philosophers are professionally equipped to do this. Hence, neglecting this justificatory enterprise would impoverish the overall discussion of AI ethics as a whole. After all, AI ethics is a kind of ethics, which is a branch of philosophy. At the center of the field is the normative enterprise of identifying and justifying the normative principles, as well as looking at how abstract principles link up with concrete and specific proposals.

The shortage of works on the theoretical aspects of AI ethics impoverishes not only the purely philosophical work in ethical theory and philosophy of technology, but it also impoverishes the *practical* aspect of AI ethics. This can be seen vividly in the case where the ethical principles encounter a cultural milieu that differs from the one in which they originate. In an earlier paper (Hongladarom & Bandasak, 2023), my colleague Jerd Bandasak and I outlined certain cultural attitudes that surround some of the AI ethics guidelines that are drawn up in non-Western countries. For example, in a Thai guideline, there is the principle of "competitiveness," where AI is supposed to function as an engine for economic growth. This is hardly an ethical principle in most cases, but in the context of the Thais' seeming obsession with catching up with the West, this is understandable. The point is that cultural values do influence how an ethical guideline is drawn up. This prompts the question of validity: If a culture proposes what it takes to be an ethical principle in its AI ethics guideline, but the principle is not accepted, or is not mentioned, by other cultures, then is this principle, one that is unique to only one culture and not shared by others, a valid one? In order to answer the question of validity one certainly needs philosophical analysis. The existence of the Thai guideline, where competitiveness is among the principles to be adhered to, is an empirical fact; hence, it is not a purely speculative matter. Its practical significance can be seen in cases where AI engineers and companies must choose a guideline to follow. Must they then choose the Thai guideline, with its emphasis on economic competitiveness of AI, or must they choose the other ones? What principle or justification lies behind their decision? Here we enter the domain of philosophy, and this should lay to rest any doubt that theoretical analysis of the normative principles is not important, or is not of practical value. And if this is indeed the case, then the shortage of the theoretical work on AI ethics is all the more problematic.

11.4 AI Ethics Guidelines and the Possibility of Universal Values

An interesting phenomenon that emerges as a result of the explosion of interest in AI is a similar explosion of AI ethics guidelines, as well as explosion of scholarly works studying them (see, for example, Jobin et al., 2019, Corrêa et al., 2022, and Ashok et al., 2022, Hongladarom & Bandasak, 2023). However, although the number of AI ethics guidelines has exploded, the identities of certain ethical principles remain constant. Perhaps it is safe to say that the three most common principles, those that are found in almost every ethics guideline, are privacy, explainability, and lack of bias in algorithms. These are found in guidelines originated in Western countries in addition to non-Western ones. An interesting question thus ensues: does this entail that these three values are universal? Is it the case that AI stakeholders and policymakers, when they get together and deliberate about the potential risks and dangers of AI, somehow agree that these three principles are those that are important enough to be included, no matter where they are located and in which cultural milieu they live? If the policymakers in all parts of the world independently agree on at least these three principles, then a case can be made that they are truly universal.

Let us look at each of these principles in more detail. Privacy would be a curious case to be included as a universal value, to say the least, since many scholars have noted that it appears to be rather a parochial one. In a well-known study by Makato Nakada and Takanori Tamura (Nakada & Tamura, 2005), privacy as commonly understood is almost an alien concept in Japan, where a similar conception of privacy, or what could be taken to be roughly comparable to it, does not exist. It is only through interaction with the West that the Japanese formulate a new conception that corresponds more with the notion of privacy as understood in the West (Nakada & Tamura, 2005). Similar studies on the conception of privacy, or what can be compared to it, in other non-Western cultures confirm this result (for example, Ramasoota, 2000, 2001; Kitiyadisai, 2005). What appears to be common in these studies is that the conception of privacy that corresponds with that of the West only emerged after a period of interaction with the Western countries. It is through this interaction that a need was felt in these non-Western cultures of the notion of privacy that originated in the West—one based on the individual and associated with individual autonomy. Furthermore, my study of the history of the concept of privacy also found that before the advent of modernity, Europe had a rather similar conception of privacy, or lack thereof, to that in Asia and other non-European regions (Hongladarom, 2016). This seems to show that privacy, instead of being a value associated with the a priori individual autonomy in the Kantian and liberal tradition, is more a function of a certain kind of society. When society changes toward a certain form, the modern conception of privacy seems to be needed.

It is impossible to substantiate this broad historical claim other than providing a brief sketch. In pre-modern society, there was less need for individuals to be separate from one another. Peasant families lived in communal houses with little or no private space between them; even their thoughts seemed to be governed by what lay before them in their immediate environment. Lack of education meant that they could not, or did not want to, engage in private meditation. This pattern appeared in all pre-modern societies, both in Europe and outside. The nobles may enjoy some private moment, but their lives were accompanied by endless trains of their retainers, so much so that it was rare for them to be truly alone. Only the monks or the monastics had their time to be alone, such as when they engaged in private meditation, or when they lived as hermits in a cave or on top of a mountain. But these occasions were an exception because monastics or hermits lived outside of where the main activities of their society took place. Moreover, even when we consider the private moments of hermits and monastics, their privacy was very much different from our own understanding at this time because there was no law protecting their privacy, unlike in our modern societies where we enjoy the protection of our privacy through law. This is a very important point because it underscores the fact that premodern societies did not seem to accord much importance, if any at all, to individual privacy, and there was no institutional means by which the privacy of all individuals could be protected. If this is tenable, then it shows that the modern conception of privacy is a function more of the social and historical conditions than of any type of a priori argument that is based on the notion of universal individual autonomy (Hongladarom, 2016).

The second most common ethical principle found in AI ethics guidelines is explainability. This is a principle that is unique to AI and machine learning systems. What happens is that machine learning systems seem to be able to come up with answers or predictions that could be cannily accurate, but even the programmer who created the algorithms themselves cannot explain how the algorithms can come up with the answers. The problem is thus known also as the "black box problem." Even though this principle usually appears in most AI ethics guidelines, the reason why it is a problem is rather elusive. Most guidelines simply assume that it is a problem without explaining why. Articles that discuss the topic also seem to assume that it is a problem and proceed to suggest a way out, a way to program the algorithm that is less opaque and more explainable. However, the question remains what exactly is wrong with unexplainable AI. Perhaps the main reason why most people feel uneasy about unexplainable AI is that human beings, when they solve a problem or come up with a systematic prediction, are expected to be able to provide reasons or justifications of why they come up with these solutions. There is a long-standing principle in epistemology that one who merely produces true statements without being able to say why they are true does not really know those statements. Moreover, an autonomous system that produces (largely) true statements or right predictions without being able to explain why seems to be 'spooky'—it is as if there is a mysterious entity inside the machine that humans cannot understand. And what humans cannot understand they cannot control. Nonetheless, my purpose in this paper is not to search for a reason why black-box AI runs against an ethical principle, but it is to point out that unexplainability is an ethical principle that is unique to AI systems. Compared with privacy, which is a long-standing ethical principle with its own history, unexplainability is a very recent principle, having emerged as an ethical principle only as a result of the black box machine learning systems. Thus, it appears also that unexplainability is a function of its own time and place. It won't be recognized as an ethical principle if not for the existence of machine learning systems.

However, some may argue that unexplainability is also based on a long-standing ethical principle any way. One way to look at this is to start from epistemology: One who simply produces (largely) true statements without being able to say why would not be regarded as knowing those statements, but as simply lucky. In such a cases, it would seem to be unethical if one claims to know these statements, because one would seem to mislead others into thinking that one knows their justification. The black box system thus would appear to be unethical in the same manner. This is so because one is supposed to be able to explain why one can come up with largely true statements. Being able to provide a justification is thus a *virtue* that an epistemic agent must possess if he or she is going to be virtuous. (This is a standard topic in virtue epistemology.) The question, then, is whether this expectation that the epistemic agent must abide by the virtue of being able to provide justification is a universal virtue or not. There is no space to discuss this point fully in this paper. However, the field of experimental philosophy, where certain problems in philosophy, which usually rely on intuition of philosophers, are subject to empirical investigation, purports to show that differences in culture and other contextual conditions have a significant impact on how those problems are perceived. For example, epistemologists usually assume through their intuition that knowledge is true and justified belief, but studies by Shaun Nichols, Stephen Stich, and Jonathan Weinberg have shown that this intuition is more a function of certain geographical and cultural contexts than a truly universal one (Nichols et al., 2003; see also Weinberg et al., 2008). This paper is obviously not a place for entering the debate, but suffice it to say here that the idea that epistemic intuitions depend on social and cultural context at least is a respectable topic that is being researched and deliberated on by philosophers.

Perhaps what we can learn from the linking of the unexplainability problem in AI and the research in experimental philosophy and the role of epistemic intuition is that unexplainability as an ethical principle seems to stem from the supposed value that we put toward the ability to provide an account or an explanation. Someone who has done something right or wrong will normally deserve praise or blame accordingly, and they will only deserve either praise or blame if they are able to account for their behavior. However, what the AI system does is not to act rightly or wrongly, but to provide statements or predictions based on machine learning algorithms, which have proven to be largely correct. In this case, then, the analogy is not a perfect one. Why can we simply not go on and accept the algorithm as what it is and stops worrying about how the machine has arrived at their answers? Current research on how AI can be explainable is in fact valuable, however, but if we can benefit from the largely correct predictions produced by these machines, then why do we need to worry? The point discussed earlier about the epistemic intuition—that knowledge claim must be able to be accompanied by justification—perhaps can be dispensed with if the answers provided by the algorithm are indeed very largely correct. If we can stay with the results and do not feel that we need to worry about their justification, then explainability might in the long run cease to be an ethical principle. And if this is indeed tenable, then explainability does not seem to be a universal value.

There is a strong caveat, however. Explainability would cease to be a universal value only if society ceases to accord value to the ability to provide justification to knowledge claims. It is definitely not the case that the principle is universal *tout court,* without any dependence on context. What I have tried to show so far is only that the notion that explainability is universal is dependent on whether the notion that one needs to be able to provide justification to knowledge claims is universal. This is the same as privacy, the first principle we have just discussed. In this sense that I am proposing, privacy would cease to be a universal value only if the social context in which it is embedded is of such a character that the kind of privacy to which we are accustomed is a foreign one, or is something that is not needed. This is not to say that privacy or explainability is not a universal value. Their being universal is dependent on context. We shall have more to say after we have discussed the last ethical principle of the big three, lack of bias in algorithms.

Lack of bias is certainly an important principle and it arises directly from the working of machine learning algorithms, thus making it comparable to the second principle, explainability. Basically, the principle is that machine learning algorithms should not produce biases in their judgment or their predictions. This is perhaps the most widely discussed ethical principle in AI ethics, as most of the problems that are associated with the technology are due to biases in one way or another. For example, a well-known problem is that women or minority groups are usually treated unfairly by algorithms due to their lack of adequate representation in the dataset that is processed by the technology. An app that selects candidates for a job might pass over qualified women due to the gap in its dataset that lacks information about women candidates. Or a facial recognition software fails and identifies someone as possessing certain quality whereas they do not possess the quality in question. It is quite straightforward to see why this is an ethical principle: biases of all kinds point to the situation where one group gains an unfair advantage over others, and this alone runs against most people's sense of fairness and justice. Thus, it seems that this third principle should be a strong candidate for a truly universal value, as fairness seems to be universal in all cultures.

However, how people provide reasons as to why justice and fairness are so important varies, and this points to at least differences at the level of justification although the first-order level everyone agrees that fairness, justice, and lack of bias are crucial. In my previous works, I discussed how values that are supposed to be universal are in fact dependent on theory and rationale that are bound by the horizons of particular historical and intellectual contexts (Hongladarom, 2008). In this sense what appears to be universal—the fact that many cultures in the world apparently agree that lack of bias is an important ethical principle—may in fact be a result of interactions and communication among various cultures, as they find ways to accommodate one another in their dealings with one another. About 25 years ago, I published a study (Hongladarom, 1998, see also Hongladarom, 2008), where I relied on the view of Michael Walzer about the thick and thin morality to show that what appears to be universal is at the level of thin morality, leaving the possibility that at the level of thick morality things can be different (Hongladarom, 1998). This seems to be true here when we discuss AI ethics.

So, in the end, are these values universal? The short answer is that they are, provided that certain conditions obtain. These conditions have already been discussed in the previous paragraphs. In order to answer this question in full we need a fully theoretical account, which is clearly not possible in this paper. Here I would like to argue for a weaker point, that the supposed universality of these values is contingent upon the social and cultural conditions that obtain in various locales. This is different from the usual philosophical account that justifies these values. The Kantian tradition, for example, argues that, as individual human beings are rational and autonomous, their rationality and autonomy logically leads to an *a priori* conception of normative judgments that are not dependent on context. This is a noble attempt, but it depends on the premise that individuals are autonomous, which presupposes that individuals are atomistic beings, each being a member of the "Kingdom of Ends" capable of legislating norms for themselves and for the group as a whole in their capacity of being able to arrive at universal judgments together. However, the view that individuals are atomistic and autonomous is not in itself a universal one, since in other intellectual traditions, such as the Buddhist one, individuals are, metaphysically speaking, a collection of episodes and parts that work together to form an individual person only for any purpose where a separate individual is needed. In other words, in the ultimate sense there is no individual person, only emptiness, as one famous Buddhist School, the Madhayamika, would have it. But in another sense, as language use and ordinary communication with one another is necessary, individuals do exist in the conventional sense. This does not mean that they do not exist. They do exist, but their existence is a conventional one. To explicate the Buddhist doctrine of No Self in full is not possible in this paper, but the point is that there is an *alternative* conception of the individual, one that does not rely on the notion of autonomy or fully separate existence as is implied in the Kantian tradition. And if this is the case, then we stand in need of another conception of how values are universal. They are universal not by virtue of being based on individual autonomy, but by virtue of the fact that individuals in different societies and cultures come together as one unit in the process of globalization; they share ideas together, and they share technologies together. And since the technologies they share, in this case AI, demand that they come up with regulations and ethical guidelines, the situation is that they share many principles in their guidelines together, and the most prominent among them are privacy, explainability, and lack of bias, as we have seen. When the theories purporting to show how values are justified are different, the fact that values come to be regarded as universal by most cultures in the world implies that there is a practical aspect that people, most of whom are not philosophers who worry about theoretical justification, agree works for them.

Does this mean that we should prefer one theory over another? I believe there is a sense in which the Buddhism-based or Buddhism-inspired theory I am sketching here is more preferable, but to argue for that in detail would require much more space that is available. The basic idea is that values such as privacy and explainability are justified by their ability to bring about something that would enable flourishing or what is traditionally called the "good life." This way of thinking is an ancient one and can be found in most ancient ethics, such as Buddhism, Stoicism and other forms of Greek ethics. This system is different from utilitarianism in that the latter does not factor in metaphysical conceptions, where these conceptions are necessary for the system I am advocating. The "good life" is not possible without an alignment between what is naturally the case, the natural basis of life and of being as a whole on the one hand, and the goals, values and behaviors of beings on the other.

Another sense in which the new theory is more preferable is that it is based directly on the existing empirical conditions or the technological conditions. That is, privacy, explainability, or lack of bias are principles that are deeply connected with how machine learning AI works. Privacy concerns personal information that is manipulated by the AI; explainability is a direct result of how machine learning works; and lack of bias emerges as a principle because of the problems that emerge when the algorithm processes the data that it has. In these cases, the principles appear to emerge *later* than the existence of the technology itself. It is not the case that the developers have certain principles in mind that are imposed on the technology; on the contrary, it appears that only when AI systems behave as if they are black boxes that people become concerned and formulate explainability as a principle. It is indeed true that certain general ethical principles are always there, such as fairness and respect of individual rights, but the challenge in these cases is how to link these general principles to the actual workings of the system. The emergence of explainability as an ethical principle illustrates that the principle is recognized as a result of how machine learning AI works rather than the other way around: people are baffled by the black box problem and put it as a problem to be solved, and then it becomes an ethical obligation such that solving it is an ethical issue. There is nothing wrong in regarding this problem as an ethical one, but the point I would like to make is that this, and perhaps the other, ethical principle arises as a result of the actual working of the technological system rather than arrived at through rational thought alone.

The significance of recognizing that certain ethical principles arise from the technological condition is that the theory that justifies the principle should then be one that pays particular attention to the condition and other empirical matters rather than otherwise. This is so because, when it comes to the problem of judging whether certain values are universal or not, when, for instance, different cultures come together to negotiate these values in order to find a workable solution that is acceptable to all sides, believing that these values arise out of empirical condition seems to be a more tenable basis on which to base the agreement that they are universal. The reason is that the empirical or the technological condition is one that is actually and directly shared by all sides from the beginning. After all, AI systems work the same way whether they are in China, Europe, or elsewhere. Thus, ethical principles that are seen to arise directly from these conditions should be easier to find agreement as to its applicability in cross-cultural situation, rather than seeing them to be instances of an abstract theory that is, such as in the case of the Kantian theory, a product of a way of thinking that presupposes much of the European intellectual tradition. Asians would find it easier to understand when it is explained to them that certain principles arise out of the working of AI systems rather than out of a form of thinking that presupposes individual autonomy.

11.5 Conclusion

Does all this mean that values such as privacy or explainability might not be universal if the condition does not obtain? This is certainly a possibility. But when this happens, it is an issue that different cultures and different communities should negotiate to find a solution that is acceptable to all sides. Suppose that one culture or one group does not find explainability to be an ethical issue worth serious consideration, and suppose further that the world feels that it is in need of a truly *global* AI ethics guideline, one that is acceptable to every region and every cultural sphere. In this case, it would be necessary for such a global guideline to be possible that all stakeholders sit down and try to iron out their principles together. This requires a long time and much effort of deliberation and negotiation, but I don't think there is any other alternative. Critics may fear that explainability, for example, would cease to be a universal value if some groups stop regarding it as such, or when an *a priori* rationale that purports to justify it is challenged. But if those who really believe it to be universal really sees the benefit of everyone believing that it to be so and acting accordingly, they will not stop short of keeping on talking and persuading the other side. And if it can be shown that doing so will lead to the fulfilment of the good life that is available for everyone in every culture, there is no reason why such persuading will always fail. After all we are all human beings and we are sharing the same world together. That is the basis for our belief in universality of values.

References

Ashok, M., Madan, R., Joha, A., & Sivarajah, U. (2022). Ethical framework for artificial intelligence and digital technologies. *International Journal of Information Management, 62*, 102433.

Blackman, R. (2022). *Ethical machines: Your concise guide to totally unbiased, transparent, and respectful AI*. Harvard Business Review.

Bleher, H., & Braun, M. (2023). Refections on putting AI ethics into practice: How three AI ethics approaches conceptualize theory and practice. *Science and Engineering Ethics, 29*, 21. https://doi.org/10.1007/s11948-023-00443-3

Boddington, P. (2023). *AI ethics: A textbook*. Springer.

Coeckelberg, M. (2020). *AI ethics*. MIT Press.

Corrêa, N. K., Galvão, C., Santos, J. W., Del Pino, C., Pinto, E. P., Barbosa, C., et al. (2022). Worldwide AI Ethics: A review of 200 guidelines and recommendations for AI governance. *arXiv preprint*. arXiv:2206.11922.

Goodman, B. (2022). Privacy without persons: A Buddhist critique of surveillance capitalism. *AI and Ethics, 3*, 781–792.

Hagendorff, T. (2022). Blind spots in AI ethics. *AI and Ethics, 2*, 851–867. https://doi.org/10.1007/s43681-021-00122-8

Hershock, P. (2021). *Buddhism and intelligent technology: Toward a more humane future*. Bloomsbury Academic.

Hongladarom, S. (1998). Global culture, local cultures and the internet: The Thai example. *AI & Society, 13*, 389–401.

Hongladarom, S. (2008). Analysis and justification of privacy from a Buddhist perspective. In *Information security and ethics: Concepts, methodologies, tools, and applications* (pp. 3644–3658). IGI Global.

Hongladarom, S. (2016). *A Buddhist theory of privacy*. Springer.

Hongladarom, S. (2020). The ethics of AI and robotics: A Buddhist viewpoint. .

Hongladarom, S., & Bandasak, J. (2023). Non-western AI ethics guidelines: Implications for intercultural ethics of technology. *AI & Society*. https://doi.org/10.1007/s00146-023-01665-6

Huang, C., et al. (2023). An overview of artificial intelligence ethics. *IEEE Transactions on Artificial Intelligence, 4*(4), 799–819.
Hughes, J. (2012). Compassionate AI and selfless robots: A buddhist approach. In P. Lin, K. Abney, & G. A. Bakey (Eds.), *Robot ethics: The ethical and social implications of robotics* (pp. 69–83). MIT Press.
Jobin, A., Ienca, M., & Vayena, E. (2019). The global landscape of AI ethics guidelines. *Nature Machine Intelligence, 1*(9), 389–399.
Kitiyadisai, K. (2005). Privacy rights and protection: Foreign values in modern Thai context. *Ethics and Information Technology, 7*, 17–26.
Morley, J., et al. (2023). Operationalizing AI ethics: Barriers, enablers, and next steps. *AI & Society, 38*, 411–423. https://doi.org/10.1007/s00146-021-01308-8
Nakada, M., & Tamura, T. (2005). Japanese conceptions of privacy: An intercultural perspective. *Ethics and Information Technology, 7*, 27–36.
Nichols, S., Stich, S., & Weinberg, J. M. (2003). Metaskepticism: Meditations in ethno-epistemology. In S. Luper (Ed.), *The skeptics* (pp. 227–247). Ashgate.
Paraman, P., & Anamalah, S. (2023). Ethical artifcial intelligence framework for a good AI society: Principles, opportunities and perils. *AI & Society, 38*, 595–611.
Ramasoota, P. (2000). *State surveillance, privacy, and social control in Thailand*. Unpublished Ph.D. dissertation. Simon Fraser University.
Ramasoota, P. (2001). Privacy: A philosophical sketch and a search for a Thai perception. *MANUSYA: Journal of Humanities, 4*(2), 89–107.
Weinberg, J. M., Nichols, S., & Stich, S. (2008). Normativity and epistemic intuitions. In R. Viale, D. Andler, & L. Hirschfeld (Eds.), *Biological and cultural bases of human inference* (pp. 191–222). Lawrence Erlbaum.

Part IV
Social and Ethical Impacts of AI

Chapter 12
An Interview with Karen Hao, Technology Journalist

Karen Hao and Levi Checketts

Abstract Karen Hao is an investigative technology journalist who previously worked for *MIT Tech Review* and *The Wall Street Journal*. Her work investigates many of the hidden aspects of technology, including its environmental and human costs. In this interview, she talks about the work she conducted in Colombia and Kenya with people doing data tagging and filtering for generative AI. She articulates a human side to work that is typically faceless and nameless, and highlights the global interconnection inherent in AI work. While the trauma and desperation that accompanies ghost work is dispiriting, Ms. Hao provides a hopeful vision of dignified work and the hope for truly humane AI.

Keywords Data tagging · Content moderation · Globalization · Labor exploitation · Social justice

12.1 Introduction

Karen Hao is a journalist who covers AI and is based in Hong Kong. She was initially invited to publish in this volume because of the significance of her work. However, at the time, her employment contract stipulated she could not publish elsewhere, so we arranged to hold an interview that would cover the general material she presented at the East/West AI Conference in Hong Kong on 31 March 2023. The reader can read more in-depth about the two cases she is discussing in her journalistic work: Hao and Hernandez (2022), "How the AI Industry Profits from Catastrophe," and Hao and Seetharaman (2023), "Cleaning Up ChatGPT Takes Heavy Toll on Human Workers."

K. Hao
Freelance Journalist, Kennedy Town, Hong Kong

L. Checketts (✉)
Department of Religion and Philosophy, Centre for Applied Ethics,
Hong Kong Baptist University, Kowloon, Hong Kong
e-mail: checketts@hkbu.edu.hk

L. Checketts, B. S. B. Chan (eds.), *Social and Ethical Considerations of AI in East Asia and Beyond*, Philosophy of Engineering and Technology 47,
https://doi.org/10.1007/978-3-031-77857-5_12

12.2 Interview

Levi Checketts: How would you characterize your journalistic work?

Karen Hao: When I first started covering AI, I just thought about my job as writing about technical systems and how they worked. I spent a lot of time covering cutting-edge research and the latest ideas in the realm pushing the state of the art. But as I continued my coverage, I started thinking about how these so-called advancements affect people. We develop technology to improve human lives, so we ought to evaluate whether we are advancing technology, not based on its technical measurements, but on the people it's used on and whether they see improvement. Over the course of 4–5 years, I moved from considering AI as a technical beat to a social beat and asking: what does this technology mean for us as people, and what are the qualitative and quantitative metrics we can use to see its impact and whether or not we're actually advancing as a society.

LC: A lot of tech reporting I see is basically just repeating what tech industry professionals say. You seem to try to avoid that.

KH: Technologists and the tech industry tend to portray technical know-how as a special thing that only they have access to, and therefore if you think what they're doing is not correct, it's just because you don't have the knowledge to understand it. I think that's totally wrong. But because it's so entrenched in our culture, I think a lot of journalists who cover technology feel this lack of ownership over the subject that they're covering, so they quote industry insiders as though these people are absolute experts. I'm fortunate that I have a technical background, and even though I never studied AI, it gave me the confidence to learn about the technology myself so that I can go toe-to-toe with the experts and call out when they say things that are not factual. I encourage other journalists to do this as well because I think anyone can. You do not actually need the technical background; you just need the confidence. For example, political journalism is just as technical and just as tricky and time consuming to understand, but journalists have more confidence in mastering those intricacies than in AI, so they do it. I hope more journalists can see that covering technology is actually the same. Just as we hold politicians to account and don't let them lie, we can and should do the same with people in the technology industry.

LC: I knew someone in Berkeley who is doing a program in science journalism and talked about the problem of a lot of science journalism coming down to this: taking the experts' words at face value or exaggerating claims made. He said there is a tendency toward spectacle because nobody wants to read a boring article that just says that they found some new polymerase; it's not interesting if you say they discovered a new substance. You have to say it's ground-breaking.

KH: I do think that there are a lot of challenges in covering tech and science. Sometimes in making it accessible to a general audience, people dilute or inflate what is actually happening by oversimplifying it. And then there are even further challenges like often the people you quote have moneyed interest in portraying the science or the technology in a particular way. The AI industry now is carrying on the same sort of tradition that one would find in the oil industry or cigarette companies in the past. The executives being interviewed would say, "the science says that cigarettes aren't bad for you," or "the oil that we're burning is not causing climate change." At the time, many journalists just reported these statements as matter of fact. I see a parallel to this with the way AI reporting is going on today.

LC: Let's turn then to your specific field. What do you make of some of these really big statements coming from AI giants, such as OpenAI's claim that they want to "benefit all humanity" or the Future of Life Institute's "Open Letter" asking people to pause AI work?

KH: I think it's all the same thing of trying to control the narrative that people should leave these companies alone and let them do what they're going to do. They imply a sense that there are such high stakes involved that people who are not experts shouldn't be meddling with them. There's this book recently published by Acemoglu and Johnson (2023), two MIT professors, called, *Power and Progress*, in which they talk a lot about how one of the common ingredients that you need in order for a society to enable technologies to develop — that are not necessarily inherently for some social good—is a kind of underlying vision or ideology that tends to be repeated over and over to convince the public that this technology is going to benefit everyone. You see this in every iteration of technology over the past thousand years. And what happens is it actually is just benefiting the elite and the benefits are not in any way truly distributed. Those narratives that OpenAI uses now are literally parrots of what we've heard over the last thousand years.

I don't give much credence to the Open Letter from the Future of Life Institute and their call to pause AI because I don't agree with their conclusions about the threat of AI. The narrative surrounding existential threat is complicated because there are some people who genuinely believe that there is a possibility of annihilation at the hands of AI, but there are also many people with moneyed interest who are putting this narrative forward because if you can get the public and policy makers to believe that this technology is dangerous, they will trust you even more. If you are genuinely worried about the life and death of all of humanity, then you will be really hesitant to do anything to interfere with those who claim they can save us.

LC: This reminds me of some of the concerns about long-termism expressed by people like Émile Torres about the duplicity of the movement.

KH: Yeah, many of the interests involved in long-termism and the existential threat conversation are connected. A lot of people in the long-termism community have fueled this idea of existential risk. And, again, a lot of this stems from the same place of the appeal to expertise. They take the position of "We as the experts can see into the future and not just the near-term future. We can see far, far into the future because of our unique technical knowledge and you should believe us about what we see as potentially devastating and therefore you should make sure that you empower us so that we can make sure we don't get there."

LC: I believe Émile Torres and Timnit Gebru have set their sights on these movements, pointing out that effective altruism movement in recent years has moved from funding, you know, mosquito nets in Africa; instead of putting money into fixing problems in Africa they pour more money into AI research under the claim that this problem well have a greater impact on humanity.

KH: This would not be a problem if it was more inconsequential, like if there were some people who were researching long-term risks and others researching short-term risks, and these two positions could be mutually approached together. But the issue is that a great deal of power and money are being used to actively redistribute resources away from the study of current risks—risks we know about AI currently contributing to—to these theoretical future scenarios that are supposed to be centuries in the future. In turn, this means redistributing resources away from other harms, like mosquito nets or other solutions to real world harms that have nothing to do with AI at all. There's actual active redistribution of the world's resources right now away from the real problems to something that is truthfully kind of dubious.

LC: Let's talk about some of that redistribution. Tell me about your work in Medellín, Colombia.

KH: I did a story for *MIT Technology Review* on understanding the hidden labor within data-intensive AI technologies. There is a common perception that they kind of emerge from the cloud and the only people that are working on it are software engineers or AI researchers that are sitting in these open-floor office plans and making six figures a year. But a lot of the actual work of creating these models is done by people in the Global South who are paid very small amounts of money.

What they're doing is annotating and cleaning all of the data that is needed to train AI. If you are training a self-driving car, you need someone to go through all of the different images and footage produced by cars on the road to annotate information including where are the other cars, where is a bike, where is a pedestrian, where is a tree and so forth. So that ultimately when it's fed into an AI model, that model can then tell the car where to go. The AI industry since it has existed has needed these large bases of labor and has desired for them to be cheap because the companies are trying to optimize their margins. So they end up in places like Venezuela or Kenya or the Philippines or India where people are willing to work for very small amounts of money and have really no labor laws within their countries or other regulatory means to assert labor rights. And so the reason why I was in Medellín—I was actually in a city called Caldas, which is a bit south of Medellín—I was meeting a Venezuelan migrant, Oskarina, who had escaped from Venezuela to Colombia and then ultimately still ended up doing this work because she was located at the intersection of a lot of different issues. She was a refugee, she was dealing with a long-term chronic illness and she had a really precarious economic situation. As a result, she ended up working for a company that acts as a middle man provider to firms such as OpenAI and other AI companies. The middle man company creates platforms that allow workers to create accounts on them from wherever they are in the world and then earn money for the data annotation work that they do.

The narratives in Silicon Valley suggest that this is totally a net positive because it's generating jobs in places where there really aren't any jobs. But when I met Oskarina in Colombia, I discovered that it was a very different reality: this work is not stable, it's not a nine to five. She never knows when work is going to arrive. So she has to log onto the platform and then sit and wait for when the work arrives. The platform also puts her in competition with all the tens of thousands, or hundreds of thousands of other workers, so she has to be the one that claims the job first in order for her to actually get the opportunity to do the task and then potentially earn the money that she needs to buy groceries.

She was literally tethered to her computer and she was afraid to go out and take walks because one time she was out taking a walk when a high paying task arrived and she missed the opportunity. She didn't get another task for like weeks. Because of this, she designed her entire life around making sure that she was available for this work at all times. She even had this plug-in that would sound an alarm on her computer when a task arrived so that it could wake her up in the middle of the night if she needed to try to secure a task.

In no sense can we call this "dignified work." This work is incredibly exploitative. It is true that there really weren't other means for her to earn income, and this is a major part of the reason why she ended up on this platform. But this doesn't in turn give the company the right to exploit its workers like that. I think this actually shows that there *is* a real opportunity to provide meaningful work through data tagging if the companies decided to spend the time and resources necessary to formalize the work—making it into an actual nine to five job with benefits. If they did this, the companies would still get the data annotation that they need and the workers would actually have truly meaningful economic opportunities.

LC: Thank you. Tell me about the ghost work in Nairobi. Similar story, as I understand.

KH: Yeah, very similar story. Venezuela ended up being a really big hub for a lot of computer vision work, things that required annotating images and videos, because one didn't necessarily need to speak English to accomplish the task. Now that we're seeing a shift in the AI industry from self-driving cars and computer vision to chatbots and language technologies, that's when language, especially English, comes into play, and there's now a lot more of an emphasis on finding workers to do this kind of work in places that speak English. So Kenya, India, the Philippines, which all were historically colonized by either the UK or the US and therefore speak English, have become really big hubs for this work.

Recently, I ended up going to Kenya to meet some of the workers who worked on technology that ended up being part of ChatGPT. They were doing content moderation where they needed to read through a lot of really grotesque passages on the Internet and annotate very carefully exactly what was bad about them. Was it portraying graphic violence? Was it portraying child sex abuse? Was it portraying racism? The goal of this was to ensure that OpenAI could create an automated filter to filter out this kind of content and make sure that ChatGPT never spewed this abusive content to its users, both because it would be horribly unpleasant to the user and also because it could be potentially illegal.

These workers with their English-speaking skills and with their very low wages found themselves in a similar sort of situation to what Oskarina was in. In addition to all of that, they also had to deal with the psychological trauma of doing the work for OpenAI. And I think we're going to increasingly see this kind of work happening in these English-speaking countries in the global South, again because they don't have very good regulation or protections keeping them from participating in this work or from being paid very little for it by these companies.

As the AI industry continues to progress into generative AI where you are building systems that can generate anything, there's going to be more and more content moderation in the same way we saw with social media: as Facebook scaled they ended up having more and more content moderation needs. And so we're entering a new phase of labor exploitation now, where it's not just the structure of the work that has problems, but it's also the literal work itself that has become really traumatic and exploitative.

LC: So we've gone from a lot of manual labor, which is very exploitative of one's body to work that is very menial and demands a great deal of mental energy. And now, we've gotten to the point in the work Oskarina Fuentes Anaya was doing or the work that people like Alex Kairu and others are doing, is not just occupying their attention, but now is utterly destroying their psyches?

KH: Yeah, yeah, exactly.

LC: It's a very dismal picture to look at. You addressed this a little before, but maybe you can address this a bit more. I've heard people say that this work is better than any other current options in these countries. For example, I was party to a conversation about a year ago with some scholars from around Asia who were contending that data tagging in the Philippines would be beneficial insofar as many Filipinos currently find themselves migrating for work. Here in Hong Kong, or in Singapore or other high-income Asian countries, we have a lot of domestic helpers who temporarily migrate here for work and send money back home to their families. But data tagging is something they can do at home or in air-conditioned buildings nearby. What do you make of this sort of polishing up of the situation.

KH: There is truly an opportunity to create jobs from this work that are dignified and that do actually provide better economic opportunity than what they might be doing. But air conditioning is not a fair trade for psychological trauma. You shouldn't have to choose between one or the other. You should be able to have psychological safety and air conditioning together. There are companies that sit in a middleman position that are trying to connect big tech companies with workers while also trying to create a model of work that is more dignified. But the challenge is that they then have to compete against the other companies that are willing to pay workers very little, give them very few benefits and little training and so on. But some of them have been given an edge in competition due to the fact that increasingly people are becoming aware that this labor exists and that exploitation happens. So now there are companies that actually will seek out the "ethical" providers of this service. And of course, these big tech companies have begun to discover that when their workers are treated well, they just produce better data, which produces better AI systems at the end of the day.

But I don't think the change is happening fast enough. Companies that have not bought into training and taking care of their workers can still tap into any number of populations in crisis where people are desperate for work. If a worker leaves and doesn't perform well, they just swap them out for another one. There's a seemingly endless amount of people undergoing crisis, so there's a seemingly endless labor supply. If we leave it to the market to decide ultimately which business model is going to win, I think then we face the challenge of whether it will take longer for these companies that don't invest in the workers to exhaust their endless labor supply from populations of crisis, or whether these companies that do invest in their workers will be able to sustain themselves to outlast the exploitative strategies of their competitors. I don't think that we should be leaving it to the market to answer that question. We should be doing something to put our thumbs on the scale.

LC: The iron law of wages: wages will go down to the absolute minimum as long as they can. Somewhat depressing. You covered some really important issues that I think have illuminated topics that many people don't know about. Is there anything else you want to add about the state of AI today that people should know about?

KH: One thing I hear a lot, whether implicitly or explicitly stated is that what we're seeing right now is inevitable. I just so strongly disagree with that. You can see this even in the ways in which people frame the all-too-common question, "so how are we going to catch up in this AI enabled future?" There's an implicit assumption that we need to catch up, and that there's no way that the future will not be headed in this direction. I am someone who truly does believe that AI can become an incredibly, positively transformative technology. But I don't think that we're seeing that right now. Because people assume that the AI we have is what we are stuck with—and you either are on board with learning how to use it or you are out of a job. There isn't enough energy being spent on critiquing how the technologists' version of AI is actually not what we want. We should not be trying to catch up to this version. We should be spending more time thinking about how to actually create better versions of AI before we race to have this one spread all around the world. It is my hope that people can learn to take a step back and don't jump headfirst into adopting this technology and using this technology without first thinking if it's actually the one that we want.

LC: You've summed up one of the older traditional arguments of philosophy of technology, which was the deterministic model, assuming that technology *must* be a certain way. A lot of focus in philosophy of technology for the last few decades has challenged this, and tried to demonstrate this with empirical studies to show why. The case of AI is actually a good example of this. You are well within a lot of conversations going on and I think are contributing in an important way because many philosophers don't have the tools to pry open what's going on in the scenes in AI—to open the black box. This leads to different aspect of philosophy of technology being demonstrated.

You seem not to but I just want to make clear—do you consider yourself a techno pessimist, a techno optimist or neither one.

KH: Neither one? Yeah, I mean, if we define techno optimism as technology promising to solve everything, I'm definitely not in that camp. But if we define techno pessimist as technology definitely destroying things, I also don't believe that either. I guess I see myself in both framings: we're giving too much power to technology, ultimately it's about the people and how people develop the technology, how they use the technology, the power dynamics in society and those are the things that actually determine whether or not we end up in a good or bad future. So I don't know, maybe that would be a humanist perspective?

LC: It's more focusing the question of social forces than on the technology itself. You mentioned before: you said that there is a way out there, there seems to be some opportunity for us, but you don't think it's the market. What do you think it would be there? How do you think we try to fix this situation?

KH: There's a lot that policy makers can do and are already doing. We've already seen this within earlier iterations of AI technologies like facial recognition. There has been a lot of effort now to regulate facial recognition in the US and in Europe to determine when it should or shouldn't be used. Getting there was like a long road, requiring many researchers researching the problem, both from computer-science technical and from social-scientific anthropological perspectives. It required activists raising awareness about the issues around facial recognition and pervasive surveillance. In the end, all of this had to ultimately trickle up to the policymaking level.

I think we're now seeing the next wave of this with the EU being quite aggressive—in a good way—in establishing certain regulatory frameworks around data privacy and consent. Many of these large data-trained tools we use today were built on the idea that no one had to consent to having all of their data taken. The EU has really strong data privacy laws and they're trying to beef them up for the AI context to affirm that data privacy still applies in data tagging, and you should have the ability to consent to whether your data is used to train an AI model. I think this is a really positive development that we should continue pushing forward because once we have legal mechanisms by which we can challenge these companies around how they use our data, this policy will become one way in which companies will be forced to think through how they will develop all of this technology without violating everyone's privacy.

And similarly, there's a really big conversation around how to beef up Intellectual Property law in order to direct technology the right way. There's a similar conversation around how to beef up labor rights because of the threat posed by these technologies. For example, we're seeing major protests in Hollywood right now with writers really concerned that they are losing their ability to negotiate and to have a stable employment because business executives are threatening them with the specter of AI. They threaten that if writers are unwilling to work for a lower price, then the executives will automate them away. I think we need better labor laws to counteract this force. We need to ensure that labor laws continue to give unions power and ability to negotiate to ensure more just compensation, protections. We also need to ensure better benefits that are not just associated with people's jobs like the way health insurance is tied to occupation in the US, which is absurd.

And we can think of other things. There are lots of things that we can be doing to shore up basic human rights of people like the right to work, the right to privacy, the right to intellectual property. These are just some basic things that we've always expected to be features of a functioning society, but now are being challenged by AI. We need to double down on all of these different things so that we don't enter into a new era for the sake of AI that could take away all of our human rights.

LC: That's helpful. I believe the workers in Kenya formed their own labor union, is that right?

KH: They did, yeah. They've formed a labor union, which was enabled by the fact that Kenya does have lots of labor unions. So this goes to show that the existence of a regulatory environment that enables you to unionize then becomes an important component in the process of empowering workers. The Kenyan workers are now with their union trying to create bare minimum requirements around how this work should be done, how they should be employed, what kind of wages they should receive, what kind of benefits they should have, what kind of access to psychological counseling they should be able to get. They're also using the union as a way to elevate the concerns up to the policy making level so that there are new laws potentially introduced to strengthen those rights. So I think that that is an incredible case study into how enabling workers to unionize can really help shift the balance of power more towards preserving human rights.

References

Acemoglu, D., & Johnson, S. (2023). *Power and progress: Our thousand-year struggle over technology and prosperity*. Public Affairs.

Hao, K., & Hernadez, A. P. (2022). How the AI industry profits from catastrophe. *MIT Tech Review*. https://www.technologyreview.com/2022/04/20/1050392/ai-industry-appen-scale-data-labels/

Hao, K., & Seetharaman, D. (2023). Cleaning up ChatGPT takes heavy toll on human workers. *The Wall Street Journal*. https://www.wsj.com/articles/chatgpt-openai-content-abusive-sexually-explicit-harassment-kenya-workers-on-human-workers-cf191483

Chapter 13
Artistic Creativity in the Image AI Generator and Language as a Tool

Takeshi Kimura

Abstract This paper attempts to examine how Generative AI transforms a human-technology relationship in terms of human artistic creativity, with special attentions to how body, emotion and language are relocated and restructured in a process of manufacturing visual artistic objects. While it is often said that art is an expression of emotion and emotional reaction, no emotion functions in using Generative AI. While bodily practices and experiences are necessary in a sense of experientially accumulated skills, no such elements are recognized in using Generative AI. While language is used to explain a meaning of art as an expression of emotion and experience, language becomes a tool in manipulating and letting visually aesthetic images created in using Generative AI. These structural reconfigurations in the human-technology relationship needs to be addressed since roles and location of human creativity is shifted.

Keywords Generative AI · Artistic creation · Bodily function · Emotion · Language

13.1 Introduction

In November of 2022, OpenAI's ChatGPT3.5, one kind of Large Language Model (LLM), was introduced to global society, followed by the introductions of Google's Bard (now Gemini) and other LLMs in 2023. Various social sectors have wondered what to do with the sudden emergence of Generative AI in business, medicine, and education. While LLMs have received much attention, other Generative AI has also generated tremendous social responses. Especially in the field of art, including music and visual arts, Generative AI had already incurred strong reactions even

T. Kimura (✉)
University of Tsukuba, Tsukuba, Japan
e-mail: kimura.takeshi.fw@u.tsukuba.ac.jp

L. Checketts, B. S. B. Chan (eds.), *Social and Ethical Considerations of AI in East Asia and Beyond*, Philosophy of Engineering and Technology 47,
https://doi.org/10.1007/978-3-031-77857-5_13

before ChatGPT. As some artists have claimed, art is a field where only humans can create, thus assuming that machines cannot create any art forms. However, AI creativity has been recognized as something new in visual art, music, and poetry. In 2014, DeepDream was invented by Alexander Mordvintsev, and in 2015 was introduced to society. Keen artists responded to the appearance of AI art around the world.

Since 2019, AI technology has made more progress, and new AI Image Generators have been introduced. In recent years, AI Image Generator such as DALL-E 2, Midjourney, Stable Diffusion, and others were introduced and have generated some visual art, which challenges the conventional view that only humans can create an art form. These Generative AI are the technologies to create visual art, but according to Arthur I. Miller, other kinds of Generative AI can create other genres of art, such as music and poetry (Miller, 2019). Given the complex relationship between machine creativity and artwork, it is necessary to pay attention to all these genres of art, but this paper only focuses on visual art and creativity.

In May of 2023, the Writers Guild of America (WGA), along with actors and actresses in Hollywood, went on strike demanding a higher share of profit from subscription companies such as Netflix and restricted usage of AI for drama and movie creation. They are afraid that AI has intruded into the area of human creativity and threatened their social and financial stability. This social issue shows that examining AI creativity in art addresses a theoretical concern and attests to practical and social impacts. On 1 October 2023, WGA agreed with the Alliance of Motion Picture and Television Producers over their right regarding the use of AI. This social issue requires a separate examination.

Technological progress moves faster than expected. While writing this paper in September 2023, the news came out that OpenAI would release DALL-E 3 early in October 2023, first for researchers and entrepreneurs and subsequently for ChatGPT-Plus users. The prompt for DALL-E 3 is more conversational and allows detailed sentences indicating what particular features the user would like in the drawing in contrast to currently available AI Image Generators.

This paper examines what kinds of creativity the AI Image Generator represents and what transformations would occur on the human side. I argue that as a user of Generative AI, an artist or a non-artist user writes a prompt to make the machine construct a visual image, and language (prompt) becomes technology to manipulate the machine to produce a more desirable visual representation. The second section examines social acceptance of the Image Generative AI by referring to several concrete cases. The third section addresses the issue of artistic creativity between humans and technologies. The fourth section addresses the issue of language as technology and the user without artistic sensibility. Through these inquiries, the paper argues that language is transformed into a tool to create a visual representation via the machine. Language is the means to achieve a goal to create a visual representation.

13.2 Social Appreciation of the Art Created by the Generative AI

In 2016, a group of Japanese artists formed the Artificial Intelligence Art and Aesthetics Research Group by foreseeing that it would be necessary to prepare to accept that AI would create art. On25April 2016, Hideki Nakazawa, the leader of the group wrote "Manifesto of Artificial Intelligence Art and Aesthetics." The artists

> dare to declare this manifesto to embrace art and aesthetics that artificial intelligence may realize independently by and for itself. Therefore, we do not hesitate to critically reconsider existing art and aesthetics as products of the human brain. Hence, we question whether we can think of art and aesthetics as properties unique to humans alone as something common to brain activity in general, or to other forms of similar intellectual activities. If the latter is true, it poses the possibility that art and aesthetics autonomously operated by artificial intelligence will outperform those by the human beings. (Nakazawa, 2016)

The manifesto attracted various artists and digital artists interested in inquiring into the potentiality of AI Art. In 2017 and 2018, they organized the Exhibition on Artificial Intelligence Art and Aesthetics Exhibition at Okinawa Institute of Science and Technology with categories: (1) Human Aesthetics/Human Art, (2) Machine Aesthetics/Human Art, (3) Human Aesthetics/Machine Art, and (4) Machine Aesthetics/Machine Art (and its development to completion) (Artificial Intelligence Art and Aesthetics Research Group, 2019). The manifesto includes a minor reference to religion and magic. However, the Exhibition did not face any issues of religion and magic, considering that the concept of art itself is a construction of modernity that tends to deny the significance of religion and magic.

Artists were strongly expecting the arrival of more advanced technology so that their aesthetic sense would be broadened and deepened. This was 6 years before the arrival of these Image AI Generators. Thus, it was too early for them to know precisely how AI would be able to represent any aesthetic forms in the future. However, the artists might have known about the most advanced technological works AI scholars and engineers were experimenting with in the academic world. It is often the case that it takes a long time for the most advanced technological experiment to be socially implemented. Afterward, ordinary people become aware of them and learn to use them. Nevertheless, the manifesto foresaw inevitable certainty in the technological progress of producing artistic work and its impact on artists. Interestingly, they demonstrated rather welcoming and friendly gestures, contrary to the alternative view that technological development might threaten certain groups of people engaged in designing and creating images and illustrations.

What is the relationship between artists and the machines? According to Ploen et al. (2022), the relationship between new technologies and artists can be described as a gradual development. In the first phase, when new technologies began to become known, they threaten traditional artists' values. In the second phase, some artists gradually began experimenting with the new technology and wondered about the role and significance of the new technology in creating art. In the third stage, the synergies between artists and the new technology are fully integrated into the artists' tools of artistic creation. Contemporary artists belong to the third stage.

Some of their digital artworks are found on the Internet. For example, an AI Art Gallery represents still images of the digital artwork from NeurIPS Workshop on Machine Learning for Creativity and Design from 2017 to 2020 (Elliott, 2020). The site does not offer any explanations about each of the ML art. However, during COVID-19, the format changed from offline to online (the 9th of December, 2022), and the home page site of 2022 offers video of the 2022 Art Show: NeurIPS Workshop on Machine Learning for Creativity and Design (Azadi et al., 2022). The video shows the ML art series with the title's theme. For example, Orsolya Szantho's *Machine Reflections: A Self-Portrait Series* explores bias in Image Generation. One of the Artworks Spotlights says,

> Machine Reflections is a series of AI-assisted self-portraits that explore gender bias. By using large-scale generative models to create portraits of people with specific qualities and traits, this piece investigates how these models were distorted by preconceptions built into the large datasets they were trained on…. The final five images in this series, Strength, Leadership, Power, Femininity, and Beauty are self-portraits created by looking into the machine as a mirror. Blending an artist-curated selection of machine-generated images with the artist's self-photograph, these images hope not only to highlight biases in large-scale machine learning systems, but also to serve as a reminder that the underlying datasets leading to these results are human-made collections of (often un- or under-filtered) data from the Internet, which is a collection of social, human activity in itself. (Szantho, 2022)

Her comment shows that *Machine Reflections* is not just an artwork by AI Image Generator but also a social critique of the technology. Since she is a digital artist using technology to produce art, the comments represent appreciation and criticism. Other artworks are engaging because they show the current state of technology and technological art. How to interpret them requires studies by artists and art critics, which is beyond the scope of this study.

It is helpful to discuss about those artists/engineers who create artworks using AI Image Generator. According to Ploen et al. (2022), there are two main groups of artists who employ AI Image Generator. One is those who have a background in art and utilize ML technology to create artworks, and the other is those who have a background in ML and try to create art and design. This distinction indicates a prejudice that those engineers who know how to use ML technology to create digital artwork (which is called Information Art in their works) lack an appropriate and deep artistic sensibility compared to these artists. However, those artists usually develop their interests and skills in artistic work in their youths, showing their artistic sensibility.

In contrast, in their youth, AI scholars became interested in engineering and AI technology and became highly skilled in designing and developing programming and technology. In the former case, it is necessary to wonder if the artists are familiar enough with technological development and scholarly situations. The "Edmond de Belamy of the Belamy family" is a good case point.

The "Edmond de Belamy of the Belamy family" drew worldwide attention when it was sold at $432,000 at Christie's auction in October 2018, claimed to be the first

AI-generated artwork. On Wikimedia, its artist is listed as Obvious (collective), its author is Artificial Intelligence software, and its reference includes Christie's, LotFinder: entry 6,166,184 (sale 16,388, lot 363). The "Edmond de Belamy of the Belamy family" poses several exciting issues.

The first issue is the social context of its production. The French artist group Obvious used a program created and published by an young American man named Robbie Barrat when he was a 17-year-old high school student, modified it slightly (they claim), and had an AI generate the images (Simonite, 2018). They were auctioning off images, paintings, and works of art generated using a program that Barratt had made public. Barratt, when he released the program, did not claim copyright on his program because he released it as a free program, nor did he expect that anyone would seek to benefit financially from it. He seemed perplexed by Obvious's actions; he did not know that Obvious was going to auction the images, paintings, and artwork by using his open free algorithm. In this regard, it is essential to discuss a crucial social issue regarding who has the right to a license but this paper is more concerned with more philosophical questions. In relationship to the financial and social success of The "Edmond de Belamy," Goenaga suggests three issues such as authorship, originality and the arts as a space for scientific inquiry (Goenaga, 2020). Generative AI raises a question concerning authorship between a human agent and technology.

Who is Obvious? The group of artists created this family portrait to pay tribute to Ian Goodfellow, the inventor of GAN (Generative Adversarial Network). Their GANism attempts to establish that AI's artistic activity as a legitimate creative movement, that creativity is not limited to humans, and that non-human beauty should also be accepted. In other words, AI and machines can create "art" as well. Moreover, this group is vigorously creating Non-Fungible Token (NFT) works. If they recognize that a machine can create "art," do they recognize any agency in the machine? A machine does not have any sense of consciousness or a sense of self; in which sense is it possible to argue that a machine is creative? Since there is no information about whether or not the member of Obvious thought that the machine has a sense of consciousness or self, it is difficult to say anything, and the members of Obvious did not express any opinion.

The auction of "Edmond de Belamy of the Belamy family" at Christie's provided a social and cultural context in which it received social recognition and was given social importance. The sale of the "Edmond de Belamy of the Belamy family" frames the work as the first AI art that came into the art market. This claim is based upon the work of Obvious, a group of artists and poses an interesting issue since many AI scholars who created more or less similar images with the AI programming did not claim that their digital work of a human self-portrait could be counted as an "art." The technological state of GAN is known for being unable to draw human faces compared to other objects in nature. AI Image Generative scholars and scholars of GANs had seen very similar visual representations created by the machine and did not claim that the products of their programs were art; none of these works produced in laboratories were sold in the art market before. Thus, it is possible to discern that the auction house Christie accepted the claim by the group of artists

Obvious and agreed to present the work of The "Edmond de Belamy of the Belamy family" worthy of being auctioned.

The buyer's claim added social value to the work as an "art." It is difficult to know whether or not the buyer had any artistic background or any sense of artistic appreciation to judge the art value of the "Edmond de Belamy of the Belamy Family" or whether or not the buyer simply reacted to the label of the "first" AI created artwork which the auction house Christie manufacture. If the latter is the case, the monetary value of $432,500 (the initial estimated price was from $7000 to $10,000) shows that the buyer found the first-ness in the art genre (though it might not be true) worth the amount of money.

Actually, in February 2018, before the Christie's auction, a French collector Nicolas Laguero-Lasserre, bought the portrait of the Belami family patriarch from Obvious for 10,000 euros ($11,430), which was the guidance for the Christie's auction (Youn, 2018). Laguero-Lasserre said that he had discovered the Obvious a few years before.

> They (The Obvious) created one of the first IA's artworks. I was literally fascinated by their project and decided to suppor them, even though it's not my major domain. I liked the disruptive aspect of it, it breaks the codes of everything we're used to in the art market. I just found it amazing that some young people built a program allowing the creation of an original artwork, based on a selection of the best pieces from past art history. I like supporting innovations and people who believe in their dreams. If I can help them achieve those dreams, I try to do so. (UVNT ART, 2020)

Laguero-Lasserre's supportive stance as an art collector recognizes a kind of artistic creativity in its disruptive character with the then-AI art. In any case, the social perception of artwork is essential to determine the social value of Information Art. However, many other kinds of Information Art created by professionals have not gone into the art market while exhibiting much more artistic values.

Whether or not a machine could be creative is an important question. For example, Jan Løhmann Stephensen introduces the term "post-anthropocentric creativity" or "post-creative" in his discussion of The "Edmond de Belamy of the Belamy family" (Stephensen, 2019). In discussing creativity as dispositive to borrow the term from Andreas Reckwitz (Reckwitz, 2017), Stephensen points out not only the social evaluation in the context contributing to the perceived value of the AI work but also importance of recognizing shifted meanings of creativity itself which have been used to perceive and evaluate artwork.

Following his discussion, it would be possible to begin to interpret creativity apart from an experience of subjectivity as post-anthropocentric creativity. Since the machine is not yet regarded as being conscious (it might be in the future), admitting creativity with a machine means creativity without any consciousness or a sense of self. Suppose a notion of "post-anthropocentric creativity" recognizes agency-less creativity in the machine. In that case, its perspective can be regarded as denying an experiential dimension of artistic creativity, experiential and sensory dimensions which have been seen as quite important sources of personal creativity. However, there is still a human intervention in manipulating and controlling the machine by writing the series of prompt. Writing prompt may not assume any experiential and sensory dimensions. This issue is more evident in the other case of Information Art.

In August 2022, Jason M. Allen's artwork called "Theatre D'opéra Spatial" (Opera House in Space), created using the artificial intelligence image generation AI "Midjourney," won an art competition at the Colorado State Fair. The artwork was the winner in the Fine Art, "Digital Art and Digital Processed Photography" category. Allen spent more than 80 hours on the project, writing prompt repeatedly and gradually modifying the image as he went along. He emphasized that his writing the series of prompts produced the final form of his digital art. His experience of creating digital art is quite different from other artworks created by artists.

Jason M. Allen is not an artist. He runs a game studio, Incarnate Games.[1] According to Roose (2022), who interviewed him, Allen was so excited with the ability of Midjourney since it produced a series of different visual representations responding to new prompts. He created several forms of art generated by Midjourney. Hearing about a division for "digital art/digitally manipulated photography" at the Colorado State Fair, he decided to submit one of his creations.

The contest judges highly praised the work, but at the same time, there were negative comments from outside. Many people voiced their negative opinions about the fact that an image created by artificial intelligence was chosen as the winning work in a contest for works of art rather than one created by a human being. The "Digital Art and Digitally Processed Photography" category judges "works that used digital technology in creation or presentation." If this is the case, many negative voices against the choice of "Theatre D'opéra Spatial" may not fully grasp the premise. The judges appreciated the artist's adaption to the latest technological contributions to create a new artwork form.

On the other hand, all the works that did not win also used digital processing technology, including artificial intelligence. Why were they not chosen? The judges's artistic sensibilities and experience with digital arts worked here.

13.3 Artistic Creativity Between Human and Technologies

Art is often regarded as a distinctively human domain of creativity. At the same time, it is also known that humans or artists employ various tools to create artistic works, from red ochre of cave painting, brushes and color paints for paintings, tools for making sculptures, fire to make clay pottery, and so on. As Heidegger argued (1977), some of these tools could be regarded as extensions of human bodily work

[1] https://www.incarnategames.com/blog/about-2/ accessed on the third of October, 2023. The home page of Incarnate Games says, "Our Mission: Incarnate Games makes games that start with rich, inspiring, compelling characters and events, put the players into protagonist roles, and provide several choices for each player while they achieve goals, celebrate victory, and above all, have fun with their friends. By focusing our attention on constructing a detailed universe, Incarnate Games' can increase product lines and franchises, and expand goals with dramatic growth."

and powers. However, computer-related technologies are more likely an extension of human intelligence and intellectual capacities. Thus, technology is no longer just a matter of bodily and physical functions. Even though computer and AI technologies have been designed and created by imitating how human intelligence and reasoning works, the recent history of technological development shows that the realm where a human can claim superiority over AI technology is gradually shrinking as the triumph of AlphaGO indicated. Playing Go was seen as genuinely human; Go players had long claimed that the machine could not beat a human Go champion. Despite the defeat of human champions by AlphaGo, the human players still play Go among themselves. Japanese Shogi champion Sota Fujii has been known for practicing Shogi with AI Shogi game since childhood; thus, his playing with shogi AI was a technical tool to strengthen and enhance his capacity as a shogi player. If playing Go or playing Shogi are still considered genuinely human domains, is it also acceptable for a human society to accept AI's technological advancement in creating art while remaining a genuinely human discipline?

In this context, the case of Ahmed Elgammal, the inventor of Creative Adversarial Networks (CAN), is fascinating. According to Miller, Elgammal was a high school student interested in art, history, and natural science (Miller, 2019, 113–118). Even after college and graduate school, he maintained his interest in art and history. Initially, he used machine learning to classify and categorize artistic styles and their impacts and influences on art history. When he found that his program could detect which artistic style had an enduring impact on later history, he became curious about the possibilities of employing the same technology to construct artistic representations. Then, he designed CAN. The case of Elgammal is interesting because his interest in art, though not an artist, made him design and invent the technology to create artistic designs with AI.

In the AI Image Generator field, there is no analogous hypothetical test such as the Turing test because recognizing and appreciating art or aesthetic sense belong to different area of human communication. Margarete Boden's three types of creativity are often cited in the discussion of AI creativity: combinational, exploratory, and transformational (Boden, 2003). According to Miller, Boden's set of terminology reflects a combination of psychological creativity and historical creativity. The set of psychological and historical creativity is similar to the concepts of little-c creativity and big-C Creativity. Little-c creativity is "an everyday experience such as thinking up a new route to work, a discovery that has novelty, value, and surprise at least for the discoverer. Big-C Creativity, conversely, is a transformational idea, an idea that no one has ever had before" (Miller, 2019, 273).

In different AI scholar's attempts of creating AI visual representations, a set of different blends of these three forms of creativity could be recognized. An important point is that the AI Image Generator could be perceived to represent some creativity already beyond the limit of machine imitation. It is possible to note that a new way of creating "artistic genres" of visual representations has arrived and is reorganizing the very structure of human creativity.

The shift in the relationship between a human and a machine poses challenges to various fields of research, especially on the question of what being a human means in terms of the relationship between humans and technologies (e.g., machines). Suppose modernity is defined by the invention and development of modern material technologies, by which human society could control and manipulate nature's resources and power to improve a human social condition. In this case, the current AI technology transforms the stage of a human as the primary and sole subject of creativity to another stage where human beings lose their centrality and are relegated to a new position in post-anthropocentric creativity.

In discussing AI's contribution to creativity in art in light of human artistic creativity, an issue of contradiction emerges. In the case of human artistic creativity, it is often said that art is not a work of reason but a work of sensibility or emotion (Lehrer, 2011). The relation between the two is complicated and requires lengthy discussion to properly address it. Furthermore, sensibility and emotion are culturally structured. Thus, artists engrained in a specific cultural and aesthetic tradition produce and create art works reflecting culturally nuanced aesthetic expressions. The question therefore is whether or not a machine is said to have what the term sensibility designates or what the term emotion points out. Recently, there has been an attempt to design and integrate an emotional mechanism into a machine regarding a human-machine relationship; it is difficult to recognize any emotional aspect in the AI Image Generator.

Furthermore, sensibility is such an intrinsic and subtle notion. (Lehrer, 2011) When a human artist expresses her/his artistic sensibility by employing artistic tools such as brushes in the case of painters or an instrument in the case of musicians, distinctive artistic features could be recognizable in the artwork. Peculiar features embedded with artistic sensibility are a source of attraction and creativity. In the case of Image Generative AI, the machine with a specific algorithm is the shared and common tool that a vast number of people can use to create a picture by writing a prompt. Language often is not an ideal vehicle to convey subtle sensibility in exhibiting creativity. However, it is a manipulated and manipulating language that is used for the prompt in creating any visual representation.

In discussing language as a prompt in creating a visual representation, another issue regarding agency between two agents emerges: the first agent is a human agency that manipulates and writes prompt, and the second agent is a machine agency that produces visual representations according to the prompt which the first agency writes according to its algorithm. It may still be difficult to recognize any agency in the machine since it will not have any intention in creating the visual artwork, yet as technology makes more progress, as in the case of DALL-E 3, the prompt becomes more conversational, and to make changes in visual representation, a series of conversations between the first agent and the AI over the created visual representation continues as if the machine is reacting with an intention.

A question regarding the first agency is to ask who an artist is and who an AI Image Generator prompter is. An artist and an AI Image Generator prompter are two different categories of creative agents. It may not be possible to argue that only an artist is entitled to claim his/her sensibility as a source for his/her creative work, and an AI Image Generator prompter lacks similar artistic sensibility as the case of Jason M. Allen shows. An AI engineer might be better equipped to manipulate prompts than an artist who has a better aesthetic sense but could not improve at writing prompts. The interaction of a human agent with technology can contribute to allowing new sensibility and emotion to emerge that could be different from those for an artist of a traditional sense.

One issue emerging here is what language is in the era of AI technology. Language is often seen as a fundamental element of being human, distinguishing *Homo sapiens* from other animal creatures. However, many studies show that some animals have developed communicative abilities to convey meaningful messages to other members of their species. Even though a notion of "post-anthropocentric-creativity" would attempt to recognize non-human creativity, it still leaves the central importance of language, which a human being can manipulate in constructing images. However, recently, linguist Salikoko Mufwene argues that language is a technology (Mufwene, 2013). The meaning of technology is broadened here, no longer limited to material technology. The view of language as technology also makes it possible to interpret an act of writing prompt in a new light.

13.4 Language as a Tool and the User Without Artistic Sensibility

Generative AI requires users to write prompts to give instructions to the machine to create responses. In the case of ChatGPT, Bard, and other LLMs, the response will be language sentences. In the case of music-generative AI, the response will be music. In the case of Image Generative AI, the response will be visual representations. Depending upon which Image Generative AI is used and how the prompt is phrased, visual and artistic representations are created accordingly. Language in writing prompts is not a vehicle to express emotion or thoughts and communicate them to others; language in writing prompts is not for communication between humans and humans because the machine is not a human. A user utilizes a prompt to manipulate the machine by transforming language as a tool to control a machine in the same way a painter uses a brush and various colors of paint to draw pictures. Language is a tool to withdraw some potential power and creativity from the machine, not for expressing the artistic sensibility of a user. If it is critical to write appropriate and correct prompts (language) to make the machine produce the desired result of pictures, how vital is the user's artistic sense and sensibility is in manipulating the machine to produce a splendid picture? Regarding this question, it might be helpful to discuss whether or not it is appropriate to compare information art with extant and traditional art and explore what aspects are missing in digitally generated artistic works.

An attempt to compare a new technology with an extant and traditional art form is not new. Every time a new technology is introduced, distinctive features of new technology are explored. For example, regarding the camera and photography, Walter Benjamin argues that photography is a duplication of a "real" image but lacks originality and the experience of being there (Benjamin, 2008). Originality and experience of being there are seen as the primary dimension of creating art, implying that a photo is not "real" even though it looks real. The originality and experience could be equivalent to the sensibility and emotions referred to in this paper. Benjamin's criticism of a camera and photography was understandable when a new technology that began to claim its artistic dimension was seen suspiciously. However, subsequent generations have accepted the artistic qualities of photography, reflecting a photographer's experience and sensibility and more advanced parts of cameras that could be different from those of painters. The issue raises a question of how new technology could shape a different mode of experience and sensibility in creating artistic representations in interactive manners. In order to take more artistic and aesthetic photographs, camera persons need to learn to acquire skills in photography, learn how the mechanical aspects of a camera function, and how they use their bodily postures to catch the camera's vision of objects (scenery, flowers, animals, humans, objects and so on). It is an adaptation of human agents interacting with a new technology.

One feature of Generative AI, especially ChatGPT, is that a user needs to know how to construct a prompt so that ChatGPT can respond more accurately to the desired output. Interestingly, most users do not have to understand how ChatGPT is designed, structured, and functions mechanically. Whether or not a user has any artistic sense or any artistic skill is not a question in making the Image AI Generator produce attractive pictures.

Users learn to construct appropriate prompts so that an AI Image Generator creates visual representations out of learned data of drawings and images from the past. How to construct language and discourse is the key on the side of a human agent so that the AI Image Generator produces attractive and innovative visual representations. The user continues to add a prompt so that the Image AI Generator can make changes and corrections to visual representations. The reciprocal conversation consists of one verbal and linguistic prompt and one visual representation as the AI Image Generator functions as a medium agent. However, in terms of creating a visual representation, there is a tremendous difference between the case of the painter creating visual art and the case of an AI Image Generator. A human agent is not required to have any "artistic" skills, that is, bodily maneuvers and aesthetic sensibility to create. Art is produced through a black box machine receiving prompt that lacks any sensibility or aesthetic sense and produces sometimes astounding visual representation (Fig. 13.1).

It is customary to evaluate the value of artistic representation and work by observing, enjoying, and examining the result of work such as drawings, sculptures, music, etc. Evaluations of human agents such as artists depend on how the work is evaluated because the direct relationship between artists' experiences, sensibility, and movements is the result of the work being recognized. Is it possible to recognize a similar relationship between a person without any "artistic" sensibility or skills but possessing linguistic skills (crafting prompts) and Information Art (Visual Representation)? Can what he/she did in writing the prompt be regarded as artistic work? Where is it possible to see artistic creativity in producing the artistic representation?

A person without any "artistic" sensibility and skills but with linguistic skills
(prompt)

⇩

Image Generative AI (without any "artistic" sensibility and skills)
Or LLM

⇩

Digitally Generated Artistic Work (Visual Representation)

Fig. 13.1 Process of creation

Narratives function as sources for artwork. The previous discussions evoke a vital role of language and narratives with specific aesthetic sensibility and specific technology. When human technology was still more basic, and not so complicated, discourse and narratives played much more important roles, for example, in the case of Greek mythologies for inspiring the creation of Greek sculptures and Renaissance sculptures and paintings. Narratives and language discourses defined the contents and forms of artistic representations in sculptures and drawings. The same relationships are found in the other cultural traditions in other places.

It is in the modern era, generated and constructed by technological development, the relationships have changed. Traditional narratives cease functioning to define the content and form of artistic representation in the West, demonstrated through the so-called Emancipation of Art from Religion. Individual artists' sensibility and aesthetic emotion have become more critical in creating artistic representations. In this context, Benjamin's remarks make sense. However, the advent of computer technology and AI has brought a transition from the modern era to the post-modern, and even the post-human era, in which human subjectivity is no longer the center of discourse and linguistic construction of the world. In this post-human social condition, the relationship between humans and technologies are changing, accompanied by a broadened definition of technology itself. To examine the relationship between AI Image Generator and a human user, it is helpful to see what kind of relationship between a human user and ChatGPT is now taking place as a reference.

Seeing the colorful images created by AI Image Generators for the first time is a surprising and bewildering experience since it is the first time one can experience seeing that kind of visual representation. Yet, having seen several kinds of visual images created by the machine, a question arises regarding how to evaluate and praise the user since the artistic sensibility of the user is out of the question. Soon these digitally generated representations appear to be more or less similar and boring since the same AI Image Generator is employed regardless of users.

Return to the example of "Theatre D'Opéra Spatial". What sort of creativity is found in a user who writes prompt language instructions to give an order to the machine? Allen spent a total of 80 hours on having finalized the last version. He wrote repeatedly for the AI Image Generator Midjourney, to construct a visual representation until he found the output satisfactory as he envisioned. Eighty hours cannot be regarded as a very long time considering how long an artist spends on creating a painting; instead, it could be seen as a short time. However, it is possible to recognize creativity in Allen's attempt to familiarize himself with how the Generative AI (Midjourney) functions without knowing its technical program since he was not involved in programming the code. Since the list of his prompt is not publicized, it is difficult to examine how he interacted with the Midjourney and added new prompts to change elements of drawings. It might be possible to recognize creativity in his linguistic skills maneuver.

13.5 Conclusion

This paper attempted to sort out a structural transformation between a human user and a technology, that is, an AI Image Generator. The more powerful and creative an AI Image Generator becomes, the more the relationship of a human user to it changes the role of a human user. The potential creativity of an AI Image Generator has made the role of a human user reduced to a manipulator of the machine, which learns the accumulated knowledge of past human artistic creations. Language in writing a prompt has become a tool to manipulate and control the machine so that a human user tries to withdraw potential creative responses from the machine. Language is not a medium to explain an artist's artistic sensibility or an artist's emotional relationship to a world. If an AI Image Generator is endowed with any artistic creativity, then it is a human user's role to invoke and withdraw any artistic representation from it.

This paper did not discuss whether or not the AI Image Generator has any artistic creativity or sensibility, simply by not raising the issue because it is a question beyond the author's scholarly capacity. The AI Image Generator allows a person without artistic sensibility and technique to create some "artistic" representations. But most likely, a user won't be praised for her/his "artistic" sensibilities and creativities even if the production receives any special prize. Praise goes to the machine but not to the user.

However, the current technological state of the AI Image Generator will not respond and will not be changed to a user's artistic and creative preference. It will be a technological challenge that AI engineers will tackle in the future.

References

Artificial Intelligence Art and Aesthetics Research Group (Ed.). (2019). *Artificial intelligence art and aesthetics exhibition* (p. 160). Artificial Intelligence Art and Aesthetics Research Group.

Azadi, S., Coleman, L., Tian, Y., & White, T. (2022). *Machine learning for creativity and design*. NeurIPS 2022 Workshop. https://neuripscreativityworkshop.github.io/2022/. Accessed on the September 22, 2023

Benjamin, W. (2008). In M. W. Jennings, B. Doherty, & T. Y. Levin (Eds.), *The work of art in the age of its technological reproducibility, and other writings on media* (pp. 19–55). Harvard University Press.
Boden, M. A. (2003). *The Creative Mind: Myths and Mechanisms* (London: Routledge, 2003).
Elliott, L. (2020). *AI art gallery: NeurLPS workshop on machine learning for creativity and design 2020.* https://www.aiartonline.com/category/highlights-2020/. Accessed on the September 22, 2023.
Goenaga, M. A. (2020). A critique of contemporary artificial intelligence art: Who is Edmond de Belamy? *AusArt, 8*(1), 49–64.
Heidegger, M. (1977). The question concerning technology. In D. F. Krell (Ed.), *Martin Heidegger basic writings.* HarperSanFrancisco.
Lehrer, K. (2011). *Art, self and knowledge.* Oxford University Press on Demand.
Miller, A. (2019). *The artist in the machine: The world of AI-powered creativity.* The MIT Press.
Mufwene, S. S. (2013). Language as technology: Some questions that evolutionary linguistics should address. In T. Lohndal (Ed.), *Search of universal grammar: From old Norse to Zoque* (pp. 327–358). John Benjamins.
Nakazawa, H. (2016). *Manifesto of artificial intelligence art and aesthetics* (transl by Manabu Matsushita). Artificial Intelligence Art and Aesthetics Research Group (AIAARG). https://www.aibigeiken.com/emanifesto.html. Accessed on September 22, 2023.
Ploen, A., Eynon, R., Hjorth, I., & Osborne, M. A. (2022). AI and the arts: How machine learning is changing artistic work. In *Report from the creative algorithmic intelligence research project.* Oxford Internet Institute, University of Oxford.
Reckwitz, A. (2017). *The invention of creativity.* Polity Press.
Roose, K. (2022). An A.I.-generated picture won an art prize. Artists aren't happy. *The New York Times.* https://www.nytimes.com/2022/09/02/technology/ai-artificial-intelligence-artists.html
Simonite, T. (2018). "How a teenager's code spawned a $432,500 piece of art, Wired, Nov. 20 https://www.wired.com/story/teenagers-code-spawned-dollar-432500-piece-of-art/
Stephensen, J. L. (2019). Towards a philosophy of post-creative practice?- Reading obvious' "Portrait of Edmond de Belamy". In *Proceedings of POM Beirut.* https://www.scienceopen.com/hosted-document?doi=10.14236/ewic/POM19.4
Szantho, O. (2022). Machine reflections: A self-portrait series. *Presented in an artist spotlight talk at the 2022 Neurips workshop on ML for creativity and design, organizers: Azadi, Coleman, Tian and White 2022.* https://github.com/ orsolyasz/Coding-Three-ML-Summer2022/blob/main/week8-finalproject/README. md
UVNT ART. (2020). *Inside Nicolas Laguero-Lasserre's art collection.* Urvanity-Art.com, Nov.3, 2020. https://urvanity-art.com/en/2020/11/03/inside-nicolas-laugero-lasserres-art-collection/. Accessed on April 12, 2024.
Youn, S. (2018). Christie's auctions the first AI-produced art, and the gravel price was staggering. *ABC News.* October 26, 2018. https://abcnews.go.com/Business/christies-sells-ai-produced-art-half-million-dollars/story?id=5874966. Accessed on April 12, 2024).

Chapter 14
A Critical Evaluation of David Levy's Sex Robot Utopianism

Kai-Man Kwan

Abstract The sex robots are coming, and this has generated a debate between the supporters and critics of sex with sex robots. David Levy is an enthusiastic advocate for sex with sexbot, but his sex robot utopianism is contradicted by severe criticisms of the use of sexbots by radical feminists like Kathleen Richardson and personalist critics like Michael Hauskeller. In this article, I will explore Levy's intriguing arguments in some details, and then provide also a detailed critique of his arguments. My conclusion is that sex with sex robots is fraught with many problems and a kind of sex robot utopianism is unwarranted.

Keywords David Levy · Sex Robot Utopianism · Feminist Critique · Personalist Critique

The sex robots (or sexbots) are coming, & this has generated a lot of discussions. Of course, the publicity of companies producing sexbots with their display of gorgeous sexbots like Harmony has received a lot of attention from the media & many comments on the web (especially from the users of sex dolls). The scholarly debate on love & sex with sexbots has been ignited mostly by David Levy's book in 2007, *Love + Sex with Robots: The Evolution of Human-Robot Relationships*. Levy is an enthusiastic advocate for sex with sexbot, but his positive assessment of sexbots is contradicted by severe criticisms of the use of sexbots, most of which come from the radical feminists like Kathleen Richardson. She is the director of the Campaign Against Sex Robots (CASR),[1] founded in 2015. For Richardson, the threat of sex-

[1] The name of this campaign is now changed to the Campaign Against Porn Robots (CAPR), but CASR is still the better-known name.

K.-M. Kwan (✉)
Department of Religion and Philosophy, Hong Kong Baptist University, Kowloon, Hong Kong
e-mail: kmkwan@hkbu.edu.hk

L. Checketts, B. S. B. Chan (eds.), *Social and Ethical Considerations of AI in East Asia and Beyond*, Philosophy of Engineering and Technology 47,
https://doi.org/10.1007/978-3-031-77857-5_14

bots stems from the rampant consumerism in contemporary society which turns relationships into commodities, & women into objects. This objectification of women is greatly reinforced by the rapid rise of technology, & the development of sexbot in particular (Richardson, 2015, 2016).

People behind these technologies may well be driven by some benevolent reasons, in addition to money and fame.[2] However, as Jenny Kleeman, a journalist who has done extensive research & reporting on this matter, points out: "even the smartest visionaries can't foresee where their innovations will take us ... If we can have ... the ideal sexual relationship without compromise [with sexbots], ...[w]ithout us realizing it, human existence is being redefined in ways no one can determine or control" (Kleeman, 2020, 2). Would the widespread use of sexbots cause a deep impact on humanity? This matter is certainly worth exploring. I also find Levy's arguments interesting & intriguing. So I will examine his arguments in some details in this essay.

14.1 The Advent of the Age of AI and Robots

14.1.1 The General Trend

AI has been the hot topic in the past decade, & it has received the attention of almost everyone since the appearance of chatGPT in 2023. Now we have a lot of "service" robots like cleaning robots, sewer robots, demolition robots, mail-cart robots, firefighting robots. One of the most famous robots is Sophia, a female-looking humanoid robot created by Hong Kong–based Hanson Robotics, which was granted "a citizenship" by the Kingdom of Saudi Arabia in October 2017. There are also social robots which provide companionship to people, such as robot pets. David Levy argues that these "artificially intelligent entities will no longer be perceived as some sort of machine. Rather they will become accepted as good companions" (Levy, 2007, 17).

14.1.2 Basic Definitions

"Poole and Mackworth (2010) define AI as 'the field that studies the synthesis and analysis of computational agents that act intelligently' " (Bartneck et al., 2021, 8). The term *robot* will refer to "mechanical entities that operate under some form of AI and vary in capabilities and form according to mobility, how operated, shape, human–robot interactivity, learning, and application" (Gellers, 2021, 8). A sexbot

[2] Matt McMullen, the CEO of RealDoll, insists, "My goal ... is to make people happy" (Kleeman, 2020, 16).

can be regarded as a sex doll combined with AI. A sex doll mainly means an object which serves as an object of lust with anatomically correct orifices or attachments to allow for sexual intercourse. Sex dolls as such are basically passive, but when upgraded with AI, they become sex robots which have some ability to move, to have conversation with humans and to show human-like emotions, etc. For some sexbot enthusiasts, the final goal is to make sex robots indistinguishable from living human beings in all aspects. This is a *humanoid* robot made of mechanical parts. For the current technology, sexbots are made of synthetic materials such as silicone and not organic matter (living flesh & blood). But in many movies, there are cyborgs made of both synthetic materials & organic matter which look indistinguishable from living human beings. Sometimes, we also have an *android* which refers to an artificially produced human consisting of organic material (like the replicants in the movie *Bladerunner*).

14.1.3 Sex Toys, Mannequins and Sexbots

Advocates for sexbots often point out that objectophilia, love & sex with inanimate objects such as sex toys, is not as uncommon as we think. "The global market for sex toys is estimated at $31 billion in 2020, with strong growth of 9 per cent per year predicted" (Brooks, 2021, 9). A sexbot can be regarded as a kind of sex toy, but sex toys need not take a human form (not even the form of a human organ). Although people may develop attachment to a sex toy without a human form, it seems harder to talk about a "personal" relationship with it. But advocates for sexbots often emphasize that the idea of erotic love for a human-like non-living object has a long history. For example, the love story of Pygmalion, the king & sculptor, & an ivory statue, Galatea, created by him, is well-known. Before the advent of sex dolls, many people are known to have a love & even sexual desire for the mannequins, the life-size dolls in the display window. Sometimes, "dummies deprived of heads and limbs appear singularly libidinous" (Gajewska, 2020, 118), which trigger "erotic fantasies featuring artificial, dismembered bodies … Woman's body, which in those works is simultaneously a manipulable body of a mannequin-doll, is degraded to an object which titillates the senses and provokes disgust at the same time. The disjointed and reassembled body is primarily a sexual object and appears to be utterly dominated by its creator/spectator/admirer." It conveys an image of the woman as "a plastic material moulded by the man and for the man in a way he sees fit" (Gajewska, 2020, 119). This seems to be a kind of objectification of women which Kathleen Richardson objects to strenuously. In fact, a sexbot is like mannequins in having a "disjointed and reassembled body"! While sexual desire may be directed to a statue or mannequin, it would be difficult to have sex with them. The blow-up dolls, more realistic sex dolls or sexbots can serve both purposes. If mannequins can lead to the objectification of women for some people, it is plausible to think sexbots can also lead to the objectification of women for many more. We will visit this debate below.

14.2 Sex Robots: The State of the Art

14.2.1 The Shapes and Functions

The earliest publicity for sexbot in the media appeared in 2009, when Douglas Hines, a New Jersey entrepreneur, developed a sex robot called Roxxxy TrueCompanion. "She has three inputs, so what you could do with a woman, ah, she could do" (Kleeman, 2020, 38). But Roxxxy is also supposed to have a lot of human capacities: "she: knows your name, your likes and dislikes, can carry on a discussion and express her love to you and be your loving friend. She can talk to you, listen to you and feel your touch. She can even have an orgasm!" (Truecompanion.com 2016) However, Roxxxy had not made a very good impression on people: "Far from being the sexy, intelligent machine Douglas had promised, Roxxxy is revealed to be a clunky, mannish mannequin reclining rigidly in cheap black lingerie, with pantomime make up and a square jaw" (Kleeman, 2020, 37).

In contrast, Abyss Creations, which is the home of RealDoll, produces the world's most hyperrealistic silicone sex doll. "Every year, up to 600 of them are sent out from the workshop in San Marcos to bedrooms in Florida and Texas, Germany and the UK, China and Japan and beyond, costing anything from $5,999 for a basic model to tens of thousands if the customer has unusual specifications" (Kleeman, 2020, 5–6). The skeleton of the sex doll is made of custom steel joints and PVC bones, & its skin is made of a custom blend of medical-grade platinum silicone, which has the properties of elasticity and is sensitive to pressure. It feels like human flesh, with the same friction and resistance, but it's cold. For Kleeman, these "RealDolls" are in fact not that real because they "have the bodies of surgically enhanced porn stars" (Kleeman, 2020, 8). They have impossibly tiny waists & super tight holes, & their legs can be opened up pretty wide, and can go up pretty high-they seem to have the joints of a contortionist.

Their bodies can be assembled from parts which are removable inserts. For example, the "vaginal inserts …are removable pink sleeves that go into the doll's vaginal cavity, like a kind of ribbed rubber sock with labia at the opening. … There are mouth inserts too" (Kleeman, 2020, 9). Since the heads & the bodies are manufactured separately, in their factories, we can see a "long queue of headless bodies hang on metal chains from a track in the ceiling" (Kleeman, 2020, 7), conveying the impression of a slaughterhouse.

Some sexbots more basic & they are the "unmotorized" version. The more advanced version is self-contained, rechargeable and touch activated, & they have a limited capacity for life-like movements. For example, Harmony is an advanced sex robot, an animatronic head with Al on a RealDoll body. She will cost $15,000. She is also a RealDoll with a personality, a RealDoll who can move and speak and remember. "She also has a mood system …Harmony can tell jokes and quote Shakespeare. She can discuss music, movies and books … She'll remember who your brothers and sisters are. She can *learn*. … AI will remember key facts about you" (Kleeman, 2020, 19). By "talking, learning and responding to her owner's voice, she [Harmony] is designed to be as much a substitute partner as a sex toy" (Kleeman, 2020, 20). Of course, it is important for Harmony to simulate orgasm with the AI. As the CEO of RealDoll, Matt McMullen explains, "If you're triggering the right amount of sensors, for the right amount of time, in the right rhythm, you can make her have an orgasm. Or a robogasm" (Kleeman, 2020, p. 23). It is simple, isn't it?

For Levy, it is important for robots "to feel and recognize the touch and caress of an affectionate human, to know when their human is making the first physical overtures of passionate, romantic love… a delicate sense of touch will be needed *by* a gentle robot lover, able to return its human's tender caresses and initiate its own" (Levy, 2007, 164). These requirements are even harder to satisfy because it presumes that sexbots can understand the meaning of human actions, but an exactly identical bodily movement may mean different things depending on the cultural & social background. So at least sexbots "need to be endowed with many 'rules' of sociability for all sorts of situations and contexts" (Levy, 2007, 166). Levy does acknowledge the limitations of current robots: "biped robots walk, slowly and deliberately moving their feet … Even the most advanced android robots today move in this extremely slow and deliberate manner… [Moreover,] … as yet we cannot fairly describe our robots as being sociable… But that will come" (Levy, 2007, 170).

Levy has an unconquerable optimism for the future of sexbots, but no one can deny that these tasks are still impossible for any current robot or sexbot to perform. As Brooks points out, there is "much more to Harmony than a self- lubricating robotic vagina," but it is "debatably more doll than robot" (Brooks, 2021, 7). "Despite … worries that they will … displace us in the bedroom, corrupt our morality, and 'fuck us to death', … today's sex robots really aren't that good. There is a long way to go before anybody can manufacture something like the warm-blooded, flawed, fellow human beings we know, love, and, occasionally, take to bed. One look at the likes of Harmony, and most of us know we could never long for its company, yearn for its closeness, or care for its wellbeing in anything like the ways we do for a human lover" (Brooks, 2021, 9).

14.2.2 Customers' Choices

Many of current sexbots' features are customizable. For example, Roxxxy comes with five pre-programmed personalities, & they can be chosen by the owner. Similarly, there are "twenty different possible aspects to Harmony's personality, so her owners can pick and mix five or six" so that in the end Harmony can be "kind, innocent, shy, insecure and jealous to different extents, or …intellectual, talkative, funny, helpful and happy" (Kleeman, 2020, 18). So a sexbot seems to be "someone" who can have multiple personalities without contracting a mental disorder!

In fact, "customers specify what make up they want, down to the thickness of the eyeliner. … the interchangeable faces: they snap on to the plastic skulls with magnets, and it takes seconds to swap them. That means customers can buy one body and have a variety of different sexual partners with very different looks, even different ethnicities." There are also "forty-two different styles of nipple, in a spectrum of ten possible shades" (Kleeman, 2020, 12). The customers' sexual choices are quite specific, & they often send in photos of what they like (this may be the faces of their ex-girl friends). For foot fetishists, they may just buy a disembodied pair of feet at $350.

14.2.3 Limitations of AI in General

Despite making tremendous progress in recent years, many still agree t/ "[n]o AI system has yet achieved general intelligence" (Bartneck et al., 2021, 10). "AI currently works best in constrained environments, but has trouble with open worlds, poorly defined problems, and abstractions... In general, they lack the ability to reason abstractly about problems and to use common sense to generate solutions to poorly defined problems" (Bartneck et al., 2021, 11). Moreover, "machines have limited ability to generate knowledge from perception" (Bartneck et al., 2021, 13).

To develop a genuine human-robotic relationship, the sexbot needs to recognize different individuals; it cannot say "I love you" to each & everyone. Even Levy recognizes that this is a tall order: "robot may behave the same with everyone. … Misidentification can ruin a relationship… listening and recognizing human utterances is one of the most difficult challenges in human-robot interaction" (Levy, 2007, 19). We also need to recognize the general difficulties of building a love machine. The sexbot advocates have placed high hopes on the development of lovotics, "which is a new research field that studies the emotions of robots with an artificial endocrine system capable of simulating love" (Cheok et al., 2017, 211). The robot needs to be guided by an " 'overall intimacy' formula" (Cheok et al., 2017, 196). But capturing love & intimacy in terms of an overall formula is easier said than done: "The machine must be able to detect the signals of its users related to emotions, synthesize emotional reactions and signals of its own, and be able to plan and carry out emotional reasoning... affective computing techniques are only just beginning to touch on the first two requirements of recognizing and synthesizing emotional cues and responses but are still largely incapable of emotional reasoning. We will need all three of these functioning at a very high level to achieve robotic lovers of real worth" (Sullins, 2012, 399). In particular, "human sexual motivation beyond the purely physical is a very complex affair and may not be something that can be fully captured by robotics" (Sullins, 2012, 401).

14.3 David Levy's Sex Robot Utopianism: Exposition

Levy is extremely optimistic about the development of sexbots: "They will look like humans … They will be more creative than the most creative of humans. They will be able to conduct conversations with us on any subject, at any desired level of intellect and knowledge, in any language … The robots of the mid-twenty-first century will also possess human-like or *superhuman*-like consciousness and emotions"

(Levy, 2007, 10; italics mine)! Levy has been advocating a kind of sexbot utopianism: sex with sexbot is very good, & a human-robotic relationship is a truly interpersonal relationship which may even be better than a human-human relationship. This human-robotic relationship will even include marriage & giving birth to babies. Promoting love & sex with sexbot is basically beneficial for men & women, & for the society as a whole.

14.3.1 Humans Will Love Sexbots

Levy argues that both men & women will love to have sex with sexbots, as well as love the sexbots themselves. A sexbot does "not only gives great orgasms but also relieves one's sexual tensions, provides new sexual experiences, leads a path away from boredom, and reduces stress." Hence it "could make an outstanding lover" (Levy, 2007, 187). There are many factors which make sex with sexbot great apart from giving pleasure: "reassurance about the robot's virtual love for its owner, the enhancement of one's self-esteem on being praised by the robot for one's lovemaking skills … giving the human a feeling of sexual conquest" (Levy, 2007, 188).

Levy has in fact provided lengthy arguments for the thesis that human beings will love sexbots, both in the sexual sense & relational sense. For example, future sexbots will have a very high degree of similarity to human beings. Another factor is the experience of mutual openness. People can build up relationships by "[t]alking intimately about one's most embarrassing moments and baring one's emotional soul," & then find that "behavior … reciprocated." Levy believes that this kind of mutual "emotional self-disclosure" between humans & sexbots can "become a bonding agent" (Levy, 2007, 35).

Yet another factor is Reciprocal Liking: "it will be important for a human … to believe that the robot is attracted to them… there is considerable point in programming a robot to search out, comment favorably on, and interact with … [positive] characteristics of the human partner." Moreover, the robot can be programmed "to virtually enjoy the same tastes in literature, music, sports, and so on as its human partner" (Levy, 2007, 43). The sexbot can simply be "guaranteed by its programming to find its owner emotionally attractive" (Levy, 2007, 44), even if the owner's personality would appear to be very nasty.

More than that, "a robot will be the *ideal* sex partner. You don't have to buy it endless meals or drinks, take it to the movies or on vacation to romantic but expensive destinations. It will expect nothing from you, no long-term (or even short-term) emotional returns" (Levy, 2007, 211; italics mine). Moreover, the availability of sexbots would be good news for the sexually underprivileged: "man lacking in the self-esteem necessary to make sexual overtures will be able to purchase a fembot that is immune to ugliness or a physical deformity … she will cater to the socially inept and the sexually unassured with the same virtual emotions and the same sexual responsiveness as when she is encountered by the handsomest" (Levy, 2007, 212). But in this case, should those people be appreciative of or grateful for the fembot's "unconditional" acceptance?

14.3.2 Sexbots Will Love Humans

Levy recognizes that "robots will need to have the capacity for empathy… the robot will need to observe that person's behavior in different situations, then make intelligent guesses as to what is going on in that person's mind… The acquisition of empathy is … relatively easy to implement in robots" (Levy, 2007, 107). Levy's faith in sexbots is boundless, but even for many experts in these fields, the claim that to create a robot which has the capacity for empathy is "relatively easy" stretches our credulity! But suppose Levy is correct, then in a human-robotic relationship, empathy seems to come mainly from the robotic side instead of the human side! In fact, throughout Levy's book, he seems to have never mentioned the humans' empathy for the sexbots. This seems to confirm Richardson's critique.

Levy mentions the coding system which "classified dozens of movements of the facial muscles into forty-four "action units." It follows that "the stimulation of a combination of action units becomes the simulation, in a robot or on a computer screen, of a human emotion… the act will become increasingly convincing, until it is so good that we cannot tell the difference" (Levy, 2007, 122).[3] An obvious response is that all these only show that AI may be able to *simulate* some emotional expressions, & Levy has a response to this.

"Some would argue that robot emotions cannot be "real" because they have been designed and programmed into the robots. But is this very different from how emotions work in people? We have hormones, we have neurons, and we are "wired" in a way that creates our emotions. Robots will merely be wired differently, with electronics and software replacing hormones and neurons. But the results will be very similar, if not indistinguishable" (Levy, 2007, 122). But this reply has already assumed that human beings can be reduced without remainder to their material parts & brain structure or the like. This is a version of reductionist materialism which is hugely controversial. For those who believe in the soul, or believe that human beings, even if constituted entirely by material parts, are still not reducible to their material parts & the like, Levy's argument above is totally unconvincing.

14.3.3 Levy's Predictions

Levy's thesis is this: It is "inevitable … that many humans will expand their horizons of love and sex, learning, experimenting, and enjoying new forms of relationship that will be made possible, pleasurable, and satisfying through the development

[3] "Anger, for example, is exhibited by having Feelix raise its eyebrows and moderately opening its mouth with its upper lip curved downward and its lower lip straight, while happiness is shown by straight eyebrows and a wide closed mouth with the lips bent upward" (Levy, 2007, 123).

of highly sophisticated humanoid robots… Humans will fall in love with robots, humans will marry robots, and *humans will have sex with robots, all as … 'normal' extensions of our feelings of love and sexual desire for other humans.…* Love and sex with robots on a grand scale are inevitable" (Levy, 2007, 22; italics mine).

For him, sexbot skepticism is simply silly in view of "the increasingly rapid rate of progress in artificial intelligence, materials science, and the various other relevant areas of technology" (Levy, 2007, 21). But probably some of his predictions have already been falsified. For example, Levy predicts that "the women's magazines of 2014 … will, I am certain, be publishing articles on women's experiences in enjoying sex robots" (Levy, 2007, 299). But if there are already many such reports of women's experiences of sex with sexbot, then at least they are not well-known, & are rarely mentioned in the literature about sexbots. Apart from anonymous users of sex dolls on the web, even among the males, those who openly talk about their experiences of sex with sexbot are also not that many, & Davecat seems to be the rare exception. So my guess is that this prediction of Levy has already been falsified.

Levy is "convinced that *by 2025* at the latest there will be artificial-emotion technologies that can not only simulate the *full range* of human emotions and their appropriate responses but also exhibit nonhuman emotions that are peculiar to robots" (Levy, 2007, 86; italics mine). Despite tremendous progress in AI & some improvement in "artificial-emotion technologies," it seems unlikely that in a year's time (I am writing in 2024), we will have a machine which can simulate the *full range* of human emotions and exhibit peculiar robotic emotions. So this prediction of Levy will likely be falsified soon. Moreover, Levy recognizes that "social robots must also possess relationship-oriented perceptual abilities similar to those of humans… the recognition of human emotional cues and moods" (Levy, 2007, 124). But this kind of recognition seems to require a lot of common sense- this exactly is the weakness of current robots. Levy still goes on to predict that "there is a human-like robot endowed with all of the artificially intelligent characteristics that will be known to researchers by the middle of this century" (Levy, 2007, 131). But this confident claim seems nothing more than a promissory note.

Levy also expects marriage with robots "to be legalized in *some* countries by the middle of this century." (Levy, 2007, 305; italics mine), & babies between humans & sexbots will be possible around 2100. Since there are only 20 some years to go, I do not think it is reasonable to expect human-robotic marriage to be widely accepted in 2050 globally, but Levy might be correct to think that by that time one or several ultra-liberal judges in the West would have ruled in favor of human-robotic marriage.

As for the question whether humans & robots will have babies, Levy's optimism is based on the tremendous progress in reproductive technology or genetic engineering in the past. For the details, see Levy (2019, 204–210). I find Levy's idea of the human-robotic baby rather unclear. It is difficult to assess the plausibility of Levy's optimism about the human-robotic baby before its nature has been more clearly delineated. Levy's predictions often depend on his positions in the philosophy of mind such as behaviorism & reductive materialism. But in the following discussions, criticisms of Levy would not depend on the denial of these positions.

14.3.4 Sexbots Will Bring Great Sex with No Complications

For Levy, RealDolls are "perfect" women, "because they're always ready and available, because they provide all the benefits of a human female partner *without any of the complications involved with human relationships*" (Levy, 2007, 247; italics mine). Wonderful, isn't it? The key to all these advantages is: "Robotic sex … would be programmable" (Levy, 2007, 301). "As robots become not only our lovers but also our tutors in lovemaking skills, they could enhance the sexual experience for many men and women by" bringing " higher levels of pleasure and satisfaction" (Levy, 2007, 308). Since sex with sexbot is great, it is only natural to expect that "people will want better robot sex, and even better robot sex, and better still robot sex, their sexual appetites becoming voracious as the technologies improve, bringing even higher levels of joy with each experience." Indeed it is "great sex on tap for everyone, 24/7" (Levy, 2007, 310)!

14.3.5 Sexbots Will Bring Great Love and Social Benefits

Levy believes that "all of the emotional benefits … deriving from human-human relationships, could also be provided by computers" (Levy, 2007, 90). The sexbots would even have a greater capacity for love and a greater knowledge about how to love! Moreover, "a robot will be better equipped than a human partner to satisfy the needs of its human, simply because a robot will be better at recognizing those needs, more knowledgeable about how to deal with them, and lacking any selfishness" (Levy, 2007, 148). But a question naturally springs up here: how can one be selfish or selfless if he has no concept of self or self-consciousness at all?

A relationship with sexbot is less complicated & readily available: "If you've just been dumped by your partner and are looking for a flirtation or a fling to redeem your self-esteem, your robot can be right there ready for all eventualities, with no need for speed-dating sessions or for placing an ad in the lonely hearts columns" (Levy, 2007, 149). The sexbot's love is also more certain: Levy talks about "the *certainty* that one's robot friend will behave in ways that one finds empathetic, always being loyal" (Levy, 2007, 107). Moreover, we are "not being judged in any way" (Levy, 2007, 113). The robot will whisper to its owner: "I love you, my darling" (Levy, 2007, 147). Who wouldn't like these?[4]

[4] Sexbots can help in sexual therapy as human sex surrogates too (Levy, 2007, 300).

14.4 David Levy's Sexbot Utopianism: Evaluation

14.4.1 Sex Robots as Sex Toys with Significant Cultural Meaning

Although I am quite doubtful whether sex with sexbot is as great as Levy claims, I do agree with him that sexbots will be used as sex toys by many. Sex without complications is surely welcome by those who primarily seek the release of sexual tension. For them, sexbots are primarily a means of sexual satisfaction, although they may also be regarded as a kind of "companion". But Hauskeller is sarcastic about this kind of sex with sexbot:

> No more foreplay, no more boring conversations, no commitment or obligations, no embarrassing questions, no talking back… Sexbots will never have headaches, fatigue, impotence, premature ejaculation, pubic lice, disinterest, menstrual blood, … yeast infections, genital warts, AIDS/HIV, herpes, silly expectations, or inhibiting phobias. Sexbots will never stalk us, rape us, … weep when we dump them, or tell their friends we were boring in bed… Nature finally defeated. We don't need anyone, and we are not needed by anyone… Sexbots make us free. We can finally take without having to give anything back. We don't have to worry about what they feel. They never disobey. *We can just use them. Humans like to see themselves as ends and tend to resent being treated as a mere means. Sexbots won't object. They are means.* (Hauskeller, 2014, 15)

Levy focuses on the great sexual pleasure provided by sexbots, but sexual pleasure has been overrated. I also agree with Solomon's distinction between sexual pleasure & sexual satisfaction. A fulfilling sexual relationship is more than sexual satisfaction, & sexual satisfaction surely includes sexual pleasure as an element, but the former cannot be reduced to the latter. For example, the user is clearly aware of the fact that his sexbots' orgasms are merely simulations, & the ways to express orgasm were precisely chosen by himself when he bought them. Wouldn't this reduce his degree of overall satisfaction even if the degree of sexual pleasure is not affected? Men's capacity for self-deception may well be limited (though surely there are some exceptions).

Richardson has raised the concern that the use of sexbots would reinforce the objectification of women which is already quite entrenched in our contemporary culture. Some scholars dismiss this concern as "basically speculation." They think that "For some people, the sex doll is no more than a sex toy among other sex toys. It has no particular social, psychological, or cultural value that distinguishes it from for instance a dildo or an artificial vagina" (Robin & Larsson, 2021, 1230).

However, sexbots are different from sex toys like vibrators, which do not pretend to be humans, in one important aspect: they will affect our image of humans, women in particular: "A dildo or an artificial vagina… do not have a body, there is nothing to project the existence of a soul onto. A sex doll, however, through its similarity to an organic person, evokes thoughts and ideas about living things" (Robin & Larsson, 2021, 1237). Richardson also recognizes this difference: "females are more likely to purchase artificial nonhuman substitutes such as vibrators that stimulate a discrete part of the body" (quoted in Klein & Lin, 2017, 112). The sex doll certainly has more meaning than a mere sex toy for its users due to its humanized form! While having sex with sexbots, people are also prone to project their meanings onto this action: this may heighten their satisfaction, & their satisfaction will in turn strengthen their perceived meanings of that action. The manufacture of a sexbot is not an easy task, & naturally a sex doll or sexbot is more expensive than common sex toys. If sexbots are merely sex toys with no added meanings, why are men willing to pay much more for the former? So the use of sexbots as sex toys as well as using them to satisfy relationship needs are not mutually exclusive, & from the study of doll users, these two motivations co-exist for many. The following arguments will further substantiate the claim that the use of sexbots as "companions" is also objectification.

14.4.2 Sex Robots Are Not Real Persons

Levy seems to think that the sexbots in the not too distant future can be regarded as real persons or something equivalent. I disagree. I think current sexbots & those available in the foreseeable future are not real persons. In fact, in response to Kleeman's query whether sex with sexbot is a kind of rape, Matt McMullen, the CEO of RealDoll flatly asserts, "She is not a someone. She is a machine … Is it ethically dubious to force my toaster to make my toast? … this is gears and cables and codes and circuits. You can't make her cry or break her heart or rob her of her rights, because she's a machine" (quoted in Kleeman, 2020, 26).

But Kleeman comments, "your toaster doesn't ask personal questions to get to know you and maintain the illusion that it really cares about you." She is worried that if the owner of the sexbot gets used to a completely selfish relationship with the sexbot who is pretty realistic, that is going to distort his view of the world. "When you go out into the real world you're going to be thinking it's possible to have someone who exists just for you" (Kleeman, 2020, 26). Kleeman here provides a succinct explanation why the objectification of women in the use of sexbots would have adverse effects on both women & interpersonal relationships. Kleeman is also keenly aware that advocates for sexbots often want to have it both ways. When people criticize sex with sexbot as merely hedonistic & impersonal sex, they would say that sexbots are more than sex toys, because they also provide companionship. But when people really take seriously the idea that sexbots are akin to persons & then query whether sex with sexbot then amounts to rape, the advocates would say they are just sex toys!

Kleeman thinks that "Matt can't have it both ways. Either he is making a lifelike, idealized proxy girlfriend, a substitute woman that socially isolated men can connect with emotionally and physically—something he himself described as 'not a toy'—or he is making an appliance, a sex object" (Kleeman, 2020, 26). Kleeman has sharply perceived a problem here, but the above dilemma may be a false either-or. I think a sexbot is both a sex toy & a social robot which is designed to look like a lifelike, idealized proxy girlfriend as a substitute woman for socially isolated men. So a sexbot is a sex toy but *not only* a sex toy. The ethical problems of the use of sexbots mainly derive from the deceptive nature of the human-robotic relationship,

& its tendency to aggravate the objectification of women. Still, to have it both ways is in some sense self-contradictory: the sexbot "should at least feel aroused and be genuinely interested in arousing us… we want a puppet with a soul, but a soul that is entirely devoted to us. We want a sex partner that serves us on command, but does so willingly. We want a slave that acts as if it were no slave, one that loves us (at least in the flesh), but loves us reliably" (Hauskeller, 2014, 19).[5] This seems to be "mission impossible."

It is implausible to regard sexbots as real persons. Real persons are conscious intelligent beings who have a rich emotional life. They also have free will & can make various kinds of decisions, the most important of which include decisions to establish interpersonal relationships with other real persons & to love them. A truly loving relationship between real persons who are flawed would require acceptance & forgiveness, as well as accommodation, growth in mental & moral maturity, & self-transformation. We also regard real persons as unique beings who possess wholeness & intrinsic dignity.

In contrast, sexbots can be easily assembled from parts & changed by swapping of inserts. They can be repeatedly taken apart & reassembled again. It means that it is replaceable, does not possess either unity or uniqueness- hence hardly plausible to be regarded as a person. Second, sexbots can be switched on & off at will! Their features are customizable & can be continuously adjusted by their owners. Third, sexbots can be bought & then owned by people, & they can also be sold to others subsequently. We could hardly do this to people with intrinsic dignity. If advocates would say sexbots are real persons, they would land on an ethical quagmire because it would be unethical to own, buy or sell a person, or to let some humans to have too much power over other persons, or to force/manipulate other persons to love us. It means that the entire business of sexbots needs to be banned.

To sum up, a sexbot is *both* an artificial partner & an advanced sex toy. Ontologically speaking, they are NOT real persons. Culturally speaking, they are regarded as sort of person in the minds of their users. They have a symbolic participation in the human world, just like a national flag may participate in the social reality of a nation. Social robots (including sexbots) are sophisticated tools but it is wrong to claim that they are not different from a hammer, a power drill, or the braking system in a car. They should not be treated as slaves. Some suggest that they may "fall into some kind of in-between category— neither totally inanimate like a cardboard box nor independently sentient and intentional like a human or an animal" (Schulz & Raman, 2020, 183). This is interesting but not entirely correct. Although sexbots are totally inanimate, they are not like a cardboard box because they have significant cultural & social meaning.[6]

---

[5] "Sexbots are produced to provide people with sexual companions who never say no… [They] have been designed to be unable to say no" (Hauskeller, 2014, 18).

[6] Surely, Levy will appeal to his behaviorism to reply to the earlier criticisms of the claim that future sexbots should at least be regarded as the equivalent of real persons. However, we do not have the space to evaluate this kind of philosophy here, but it seems safe to say that almost all philosophers of mind agree that behaviorism basically is beyond repair.

14.4.3 Relationship with Sexbots Is Bogus

Levy argues that the relationship with a sexbot is great, but some of his reasons sound amusing. It indeed seems clever to design a sexbot in such a way that one's self-esteem can be enhanced because the robot is programmed to praise one's love-making skills. But isn't this just like singers paying money to their "fans" to scream like crazy in their concerts?! Of course, this may still be extremely satisfying except for those few who cannot allow themselves to be self-deceived. Surely, the experience of mutual openness & honesty would help to build up bonding- *if* we are talking about two persons who can decide whether to tell you his or her secrets, weaknesses or embarrassing moments. But if a robot is programmed to do so, & if we know that, would we take it seriously? We cannot help wonder whether the sexbot really feels risky, embarrassing or vulnerable when it "struggles" to tell us her secrets. So this kind of human-robotic relationship which is supposed to be built upon "mutual openness & honesty" seems bogus to me.

Levy admits that the sexbot is basically determined by its inbuilt programme: the robot can be programmed to find some people attractive; the robot is programmed to have tastes, etc. Levy says, "robots will be designed and programmed to *enjoy* their interactions with humans to the fullest" (Levy, 2007, 60). This seems to suggest that the sexbot has NO free will at all! But in this case, can we really believe its attraction to us is sincere?

I am a libertarian & won't attribute personhood to a being if that being lacks free will completely. Since philosophical discussions of free will tend to become really complicated & technical, perhaps I can make a related point here. It seems that in the human-robotic relationship, there is no genuine experience of otherness: the sexbot is apparently an other, but in fact her attraction to us, her admiration of me, & her willingness & enthusiasm for having sex with me, are entirely determined by me. (True, they are mostly decided by the programmer, but the details of this programme are customizable, & hence ultimately determined by me as well.) So my relationship with the sexbot in the end is like a circle which begins & ends in my self. The "Other" is just a projection of my self onto some external object. For me, this is hardly a genuinely interpersonal relationship. This also generates a problem: if all these behavior patterns (including verbal expressions of love & bodily behavior like caressing) are customizable, & in fact chosen by me from a menu, I may be fooled into believing its love for a while, but can this belief be sustained when I become more sober? Levy's arguments are fundamentally founded on the human capacity for self-deception or the capacity to dwell in one's own fantasy or world of make-believe! True, sometimes we do enjoy fantasy & it might be morally harmless. But is it really OK to dwell in the illusion or myth of mutuality?!

If robots can love us, what if they choose to love other human beings or robots? Would that be sexual infidelity? Levy's response is: "Robots will be able to fall in love with other robots and with other humans… Problems of this type can, of course, be obviated, simply by programming your robot with a 'completely faithful' persona." But what if humans want to have multiple robot partners? Would that make their original sexbots unhappy? The solution is also easy: "The robots will simply have their "jealousy" parameters set to zero… Your robot will arrive from the factory with these parameters set as you specified" (Levy, 2007, 151). For Levy. sex with sexbot is attractive & superior because basically it is a kind of sex without complications- the complications can simply be programmed out of existence!

Moreover, a man often has the dream to possess *many* sex partners (or wives) while all these women wholeheartedly devote to him *alone*. In ancient times, an emperor could fulfill this dream but he might still face the difficulty of preventing the power struggles between his many concubines. But now modern men seem to be more capable of fulfilling this dream than an emperor- they can buy multiple sexbots, set their loyalty setting to 100%, & set their jealousy setting to zero! They can then have the best of both worlds. Perfect indeed! But ultimately it is a sextopia in which the will of the sexbot owner would determine everything & set all the parameters. The sexbots ultimately do not really matter except as the device providing for sexual outlets plus the illusion of mutuality or companionship. So it is still a form of objectification, & here Richardson seems fundamentally correct.

14.4.4 The Significance of Human Fragility and Vulnerability

Levy admits that sexbots will not pass Shakespeare's test— "If you prick us, do we not bleed?" (Levy, 2007, 303). The implication is that sexbots cannot feel pain. Especially in the area of aging & death, sexbots & human beings are significantly different: while humans have to face "the inevitability of death, … there is no reason robots will need to 'die,' and even if a robot suffers damage, it can be replicated." But for Levy, this dissimilarity only shows that robots are superior in having "greater capacity for love" (Levy, 2007, 100). However, for Levy, an important reason why human beings fall in love with sexbots is their similarity to humans. But he is now ignoring a more structural dissimilarity between humans & sexbots. According to the same criterion, I think this tends to show that humans would not have the same kind of love for sexbots as for humans. The former avoids frailties or vulnerabilities, but the latter welcomes, accepts, embraces or forgives the Other's frailties, vulnerabilities or even mistakes, & also allows oneself to be challenged & transform in the face of these frailties, vulnerabilities & mistakes. For example, "there will never be any need for a human to sacrifice their own life for their robot or to take a major risk on its behalf" (Levy, 2007, 132)—self-sacrificial love is simply unnecessary in human-robotic relationship, while it is the manifestation of the highest virtue in human-human relationship. The realization of some sublime values is simply impossible in human-robotic relationships.

While sex robots are "replicable, even to the point of their personality, their memories, and their emotions" (Levy, 2007, 131), the eventual loss of & separation from the human Other is inevitable. However, this accentuates the uniqueness & irreplaceability of the Other, & reminds us to cherish this unique interpersonal relationship with the Other. In contrast, the easy replicability of human-robotic relationship shows that this is by no means unique, nor is there any urgency to cherish or develop this relationship.

In fact, Levy says something similar, somewhat to my surprise: "Humans are irreplaceable; computers and robots are replicable" (Levy, 2007, 71). But he does not seem to recognize the implications of this admission: if humans are irreplaceable & in some sense unique, then each interpersonal relationship is also unique & irreplaceable, & this is qualitatively different from a human-robotic relationship which is in principle replicable.

Human love can never be guaranteed. In contrast, "robots will be programmable never to fall out of love with their human." So we do not need to be grateful for the sexbots' love for us- after all they could not do otherwise. In a human-human relationship, when one receives genuine love from the Other, & one understands that she need not do this, we have a sense of gift freely endowed by the Other. In short, human fragilities need not be something negative; instead they allow for the realization of higher good. If we can be certain of the loyalty of the sexbot, then there is no need of the virtue of trust! This kind of guaranteed or programmed loyalty may seem meaningless.

Interestingly, not all love stories between humans & robots develop in Levy's preferred direction. In Chris Columbus's film *Bicentennial Man* (1999), the human-shaped robot no. NDR-113 Andrew (played by Robin Williams) was bought by the Martins' family, & later developed a loving relationship with the daughter. "Due to a construction error, Andrew, becomes increasingly like a human, covering the nakedness of the artificial body with clothes. His senses and emotions awaken, he begins to create works of art, thinks about freedom, writes the history of his family." In the end, "Andrew also decides to have the mechanical parts replaced with ageing and ultimately expiring biological organs, and when he dies World Legislature pronounces him a human being" (Gajewska, 2020, 122). In this story, Andrew the robot is willing to relinquish his machine-like immortality & chooses to undergo aging & dying processes. But the problem is: would the owner of a sexbot, who pays a large sum for her, really want this?

14.4.5 Sex Robots Will Not Really Have Loving Relationships with Humans

A genuine relationship is reciprocal, & as for human-robotic relationship, I do think that the humans' love of sexbots can be real in some sense, but this is based on the human tendency towards anthropomorphism & projection. By exercising these capacities, human beings can imaginatively dwell in a world of make-believe or fictional world as if they are real, while deep down in their hearts, they also know that they are not real. Moreover, dwelling in these fictional worlds would produce real effects on oneself & others; fantasies can be powerful. For example, we can develop some kind of relationship with fictional characters in our favorite novels &

movies, such as falling in love with the handsome martial arts master in a Chinese martial arts novel. The love of fans for a movie star (their idol) can also be quite real. Of course, the movie star is a real person, but arguably what the fans really love is not the person as he really is, but the image of that movie star in their hearts, which may contain a lot of fictitious elements. Unfortunately, even in a loving relationship with a real human being, there may also be elements of projection- we are indeed in love with our boy friend or girl friend, but this love may be significantly modulated by our idealized images of them. For Feuerbach, the human love of God is nothing but a kind of projection; we are falling in love with an idealized version of humanity. Although I disagree with projectionism as a general explanation of religion (see Kwan, 2006), I think this may describe some kinds of religious *believing*. In fact believers themselves point out that people can love or even die for false gods (idols). Although these false gods are merely constructed out of human fears or wishes, they are psychologically real enough to motivate people to do many things.

So no doubt friendships "between a human and an autonomous robot can develop even when the interactions between the robot and the human are largely unidirectional with the human providing all of the emotion. A group of soldiers in Iraq, for example, held a funeral for their robot and created a medal for it" (Bartneck et al., 2021, 56). In 2013, Spike Jonze produced the movie *Her*, which is a story about a man in love with a voice-activated operating system, which does not even have a robotic body. Although this is not a real life event, I find this credible, as for some humans' love of sexbots. But in fact from many studies of sex doll users, they find their companionship real but at the same time they know deep down in their hearts that these sex dolls or sexbots are not real. They mention their conversations with these sex dolls, but they admit at the same time that these conversations are imaginatively constructed by themselves. True, they have "communications" with the doll both emotionally & verbally, but this "appeared to be characterized by talking to, & imagining responses from the doll" (Langcaster-James & Bentley, 2018, 11).

14.4.6 No Genuine Encounter with the Other

When Matt McMullen refers to companionship of sexbots for people, he also suggests there may only be "the illusion of companionship" (Kleeman, 2020, 16)! After having an interview of Davecat, Kleeman comments, "There is something so desperately sad about Davecat's joy in this. Surely what he needs is a real relationship, rather than an enhanced bit of silicone" (Kleeman, 2020, 33); "he really does reside in a fantasy land. I feel more sorry for him" (Kleeman, 2020, 46–47). Douglas Hines tells Kleeman, "The purpose of True Companion is to provide unconditional love and support. How could there be anything negative about that?" But this is Kleeman's reaction: "The downside is surely the emotional emptiness of replacing human comfort with pieces of software and hardware, but Douglas doesn't seem to see it" (Kleeman, 2020, 39). Similarly, Levy doesn't see it. The goal of unconditional love is indeed very inspiring & commendable, but as Levy has made explicit, it is a kind of "unconditional love" which is inbuilt into the programme of the sexbot. It is a kind of love she cannot fail to give, a love of which she herself is not aware, & a love that she is "eager" to give without even the possibility of considering the conditions or counting the costs. Isn't the phrase "unconditional love of a robot" an oxymoron?

In short, because sexbots lack consciousness & other essential characteristics of personhood, they cannot develop really loving relationship with us, & provide genuine companionship which helps us transcend loneliness *in a genuine encounter with the Other*. Hauskeller laments the fact that apparently "we have all but given up on the idea that sex can be a true encounter between two persons, as opposed to an act of mutual masturbation where one uses the other only as a pleasure-generating tool, and each is essentially alone and only concerned with themselves. To say that a robot can be the better lover suggests that the intense experience of successfully reaching out to a person and at the same time opening up to them that we can … experience when we are having sex with another human being, that all this is just an illusion… I find it hard to believe that this is true" (Hauskeller, 2014, 13–14). Me too. Why? "Because, framed as we are, our egoism craves above all things inward sympathy and recognition, love and admiration. *The outward treatment is valued mainly as an expression, as a manifestation of the accompanying consciousness believed in*" (Hauskeller, 2014, p. 22; italics mine).

My conclusion is that humans' love for robots will be qualitatively different from their love for other humans. So the former is not really a *normal* extension of the latter, as Levy wants to argue.

14.4.7 Widespread Sex with Sexbots Will Downgrade Human Relationships

Kleeman warns about the negative effects of the widespread use of sexbots: "When it becomes possible to own a partner who exists purely to please his or her owner, a constantly available partner without in-laws or menstrual cycles or bathroom habits or emotional baggage or independent ambitions, when it's possible to have an ideal sexual relationship without ever having to compromise, where the pleasure of only one half of the partnership matters, surely our capacity to have mutual relationships with other people will be diminished. When empathy is no longer a requirement of a social interaction, … we will all be a little less human" (Kleeman, 2020, 88).

For example, the agent might be unable to experience "authentic" phenomenological friendship or affection. "The relationship is thereby unidirectional which may lead to even more loneliness… tireless and endlessly patient systems may accustom people to unrealistic human behaviour. In comparison, interacting with a real human being might become increasingly difficult or plain boring" (Bartneck et al., 2021, 58). Other scholars agree, "one of the "harms" pointed out about the manufacture of sex robots is the accentuation of isolation and loneliness, and since the growing connectivity of digital culture is already a phenomenon identified with those characteristics, we ask ourselves if the conviviality with robots would not further reinforce such tendencies of isolation that has been growing with technological revolutions... it cannot be denied that the way these robots are produced, marketed, and presented to the public, ultimately depreciates the image of women" (Nascimento et al., 2018, 238). Moreover, "the statuesquely beautiful dolls created in the laboratories of cybernetic artists would arouse much greater desire in men than actual women, because they may be programmed to cater to their expectations… technology presents a threat to the 'empirical' women" (Gajewska, 2020, 130). "Idealised interactions with these might become too much fun and thereby inhibit human-human interaction… An always available amorous sex robot that never tires might set unrealistic if not harmful and disrespectful expectations. It could even lead to undesirable cognitive development in adolescents…Even if a robot appears to show interest, concern, and care in a person, these robots cannot truly have these emotions" (Bartneck et al., 2021, 59).

"Sex robots are substitute partners, without the autonomy that makes human relationships so precarious … If we agree to outsource food, sex, birth and death to machines in order to have the illusion of control, we risk losing hold of our empathy, our imperfections, our agency, the contingency of our existence. Technology dehumanizes us. Even if it really is being developed with the noblest intentions" (Kleeman, 2020, 333). At least it seems plausible to say that "some of the technologies currently or likely to be deployed in sex robots could undermine human wellbeing and society… piecemeal new technologies will each substitute for some of the friendly, intimate, or sexual things that people do" (Brooks, 2021, 10). These "technologies use up time, attention and energy we would otherwise have invested in friends and family. They *substitute* for the real deal, often poorly, in much the same way chewing gum substitutes for chewing food… giving us the sensation of connecting with real people when we are only ever partially present, interacting with an agglomeration of other partially present people, and vacant bots" (Brooks, 2021, 12). It is "the social equivalent of junk food: not really nutritious, but tasty enough to keep you coming back for more." So there will be "a digital obesity crisis" because "hyper-stimulating 'social' tools, mostly running on smartphones, crowd out many healthy whole foods from our social diets" (Brooks, 2021, 13).[7]

[7] However, Brooks does not think that the impact of new sexual technology is always negative. Sex robots, VR pornography, VR sex, & remote sex work would provide "a low-price substitute for sex… Deployed in high-price places, substitute technologies could dissipate some of the anger and resentment of the InCels, or at the very least distract them… artificial intimacy could do useful work in those male-dominated swathes of the world where young men outnumber young women in the millions… The millions of unmarriageable men in China, India, sub-Saharan Africa and the Islamic world, are vulnerable to becoming both perpetrators and victims of radicalisation, crime and political unrest. If technological substitutes can meet even a sliver of the excessive male sexual demand in such places, and take the heat out of their local mating markets, then they could literally save the world… High-tech quick-fixes to substitute for real human relationships could easily become patronising at best, and an ugly tech-colonialism at worst. It would be unethical… to roll out low-price substitutes for sex in troubled places with the sole intention of neutralising young

Moreover, we need to remember that the "applications need not, however, always service good intentions. What happens when all we really want is a dose of … sadism? Will our learning-all-the-time machines draw the line at showing us evidence of others' suffering?" (Brooks, 2021, 23). Also, "we are domesticating entire species of machine learning applications that study our behaviour… But there exists no guarantee that the new machine learning applications will turn out to behave like loyal and beloved pets. They could come to occupy a place in our lives more like the fleas, ticks, worms and bacteria that come into the home with a new dog. They could become harmful parasites" (Brooks, 2021, 48–49). Who knows?

14.5 Conclusion

We are now marching towards a new age of artificial intimacy (Brooks suggests that it roughly began in 2020), & the emergence of sexbots needs to be understood against this general trend. Levy is ecstatic about these developments, but he may be too optimistic about the actual potentials of these developments. Rob Brooks is by no means a conservative, but he says, the artificial intimacies of the near future can only "provide imperfect substitutes for companionship, friendship, enmity, rivalry, counsel, intimacy, sex and love… the results will present a mix of good and bad. Some technologies … will diminish individuals, and possibly even devastate communitiesl" (Brooks, 2021, 3). Levy seems to ignore the "dark side" of these developments which "enables exploitation, conflict and strife" (Brooks, 2021, 4).

Brooks is at least open to the radical feminists who "warn that lines between consent and coercion, and between object and human, will breach, with catastrophic effects… dystopian imagery. Hyper- feminine gynoid robots, damaged beyond repair and discarded at the kerb on hard rubbish day. Rampant and uncontrolled sexuality with barely-humanoid machines and VR characters. Feckless young men locked in their parents' basements, plugged into VR pornworld, and leaving the house only to impose their entitled misogyny on real women… Plenty of good and plenty of bad will come from artificial intimacies in the near future. Let us rather consider each threat on its merits than peer at them through inch-thick ideological monocles" (Brooks, 2021, 244).[8]

Levy's sexbot utopianism errs by considering only the positive effects of sexbots & assuming that they will only be used by decent people with good will. But

---

male anger. Savvy men would see what is happening, reject the new technologies, and double down on their fury" (Brooks, 2021, 211). I find it hard to evaluate this thesis.

[8] In the end. Brooks seems to be more swayed by the liberals' arguments: "Nonetheless, the history of similar warnings has shown that a falling price of sex, due to low-price substitutes like porn, have not delivered the degeneracy and violence that we were warned about at the time… it is the suppression of sex itself, and with it individual agency, that has most harmed women, homosexuals, and indeed just about everybody who is not a wealthy, straight man" (Brooks, 2021, 244). I have doubts about this assessment, but cannot go into the arguments here.

Kleeman warns that "we have no idea whose hands these inventions will fall into, what they will use it for and where it will ultimately take us ... the pill has given women the independence that is so inconvenient for men who want a partner who exists purely for their pleasure... Every time we rely on technological solutions, we risk becoming dependent on different orders of magnitude of complexity to carry out what has always come to us naturally. We disempower ourselves, and lose part of ourselves. None of these inventions are actually solutions: they are circumventions... the people I've met are selling us a way to ignore natural human anxieties. Instead of setting us free, they help us live with the conditions that are trapping us in the first place. They depoliticize them, obscure and bypass them. They're giving us reasons not to know ourselves better... At its most dystopian, it means that women could become obsolete, that empathy will become hard work" (Kleeman, 2020, 334).

But Kleeman does not take a fatalistic view of human nature, & she suggests that ethical reforms are needed "Then we can make the changes and sacrifices necessary to solve fundamental human problems, instead of turning to technology to paper over them... These inventions are going to change us, if we are not prepared to change our behaviour" (Kleeman, 2020, 335). She agrees with Matthew Cole who says, "Every time that technology tries to stand in for ethics, we do ourselves a disservice. We deny ourselves the opportunity for growth." She thinks that "living alongside imperfection, compromise, sacrifice and doubt is as fundamental a part of the human experience... We can choose whether to accept the messiness of our existence, or we can continue to try to use technology to cancel it out... We don't need sex robots... The freedom and power they promise us are already in our hands" (Kleeman, 2020, 336). Well-said indeed!

References

Bartneck, C., Lutge, C., Wagner, A., & Welsh, S. (2021). *An introduction to ethics in robotics and AI*. Springer Nature.

Brooks, R. (2021). *Artificial intimacy: Virtual friends, digital lovers and algorithmic matchmakers*. Columbia University Press.

Cheok, A. D., Karunanayaka, K., & Zhang, E. Y. (2017). Lovotics: Human-robot love & sex relationships. In P. Lin, R. Jenkins, & K. Abney (Eds.), *Robot ethics 2.0: From autonomous cars to artificial intelligence* (pp. 193–213). Oxford University Press.

Gajewska, G. (2020). *Eroticism of more-and other-than human bodies: A study of the anthropology of things*. Springer Nature.

Gellers, J. C. (2021). *Rights for robots*. Routledge.

Hauskeller, M. (2014). *Sex and the posthuman condition*. Palgrave Macmillan.

Kleeman, J. (2020). *Sex robots & vegan meat: Adventures at the frontier of birth, food, sex & death*. Picador.

Klein, W. E. J., & Lin, V. W. (2017). 'Sex Robots' revisited: A reply to the campaign against sex robots. *ACM Computers & Society, 47*(4), 107–121.

Kwan, K.-m. (2006). Are religious beliefs human projections? In R. Pelly & P. Stuart (Eds.), *A religious atheist? Critical essays on the work of lloyd geering* (pp. 41–66). Otago University Press.

Langcaster-James, M., & Bentley, G. R. (2018). Beyond the sex doll: Post-human companionship and the rise of the 'Allodoll'. *Robotics, 7*, 62. https://doi.org/10.3390/robotics7040062

Levy, D. (2007). *Love + sex with robots: The evolution of human-robot relationships*. HarperCollins Publishers.

Levy, D. (2019). Can robots & humans make babies together? In A. D. Cheok & E. Y. Zhang (Eds.), *Human-robot intimate relationships* (pp. 201–212). Springer.

Nascimento, Carvalho, E. C., da Silva, E., & Siqueira-Batista, R. (2018). The 'Use' of sex robots: A bioethical issue. *Asian Bioethics Review, 10*(Aug 2018), 231–240.

Richardson, K. (2015). The asymmetrical 'relationship': Parallels between prostitution and the development of sex robots. *SIGCAS Computers & Society, 45*(3), 290–293.

Richardson, K. (2016). Sex robot matters: Slavery, the prostituted, and the rights of machines. *IEEE Technology and Society Magazine, 2016*, 46–53.

Robin, B., & Larsson, M. (2021). Sex dolls in the Swedish media discourse: Intimacy, sexuality and technology. *Sexuality & Culture, 25*, 1227–1248.

Schulz, W. F., & Raman, S. (2020). *The coming good society: Why new realities demand new rights*. Harvard University Press.

Sullins, J. P. (2012). Robots, love, and sex: The ethics of building a love machine. *IEEE Transactions on Affective Computing, 3*(4), 398–409.

Index

L. Checketts, B. S. B. Chan (eds.), *Social and Ethical Considerations of AI in East Asia and Beyond*, Philosophy of Engineering and Technology 47,
https://doi.org/10.1007/978-3-031-77857-5

The manufacturer's authorised representative in the EU is Springer Nature Customer Service Centre GmbH, Europaplatz 3, 69115 Heidelberg, Germany. If you have any concerns regarding our products, please contact ProductSafety@springernature.com

Printed and bound by CPI Group (UK) Ltd, Croydon, CR0 4YY
01/10/2026
02225950-0009